Lightning Science and Lightning Protection

Some Selected Topics

About The Centre

The Centre for Science and Technology of the Non-Aligned and Other Developing Countries (NAM S&T Centre) is an inter-governmental organisation with a membership of 47 countries spread over Asia, Africa, Middle East and Latin America. Besides this, 12 S&T agencies and academic / research institutions of Bolivia, Botswana, Brazil, India, Nigeria and Turkey are the members of the S&T-Industry Network of the Centre. The Centre was set up in 1989 to promote South-South cooperation through mutually beneficial partnerships among scientists and technologists and scientific organisations in developing countries. It implements a variety of programmes including international workshops, meetings, roundtables, training courses and collaborative projects and brings out scientific publications, including a quarterly Newsletter. It is also implementing 5 Fellowship schemes, namely, NAM S&T Centre Research Fellowship, Joint NAM S&T Centre – ICCBS Karachi Fellowship, Joint CSIR / CFTRI (Diamond Jubilee) - NAM S&T Centre Fellowship, Joint NAM S&T Centre – ZMT Bremen Fellowship and Research Training Fellowship for Developing Country Scientists (RTF-DCS) in Indian institutions. These activities provide, among others, the opportunity for scientist-to-scientist contact and interaction, training and expert assistance, familiarising the scientific community on the latest developments and techniques in the subject areas, and identification of technologies for transfer between member countries. The Centre has so far brought out 62 publications and has organised 94 international workshops and training programmes.

For further details, please visit www.namstct.org or write to the Director, NAM S&T Centre, Core 6A, 2nd Floor, India Habitat Centre, Lodhi Road, New Delhi-110003, India (Phone: +91-11-24645134/ 24644974; Fax: +91-11-24644973; E-mail: namstcentre@gmail.com; namstct@bol.net.in).

Lightning Science and Lightning Protection

Some Selected Topics

— *Editors* —

Estelle Trengove

Foster Chileshe Lubasi

CENTRE FOR SCIENCE & TECHNOLOGY OF THE NON-ALIGNED AND OTHER DEVELOPING COUNTRIES (NAM S&T CENTRE)

2015

DAYA PUBLISHING HOUSE®

A Division of

ASTRAL INTERNATIONAL PVT. LTD.

New Delhi – 110 002

Cataloging in Publication Data--DK
Courtesy: D.K. Agencies (P) Ltd. <docinfo@dkagencies.com>

African Regional Training Programme on Lightning Protection (2013 : Kampala, Uganda)
Lightning science and lightning protection : some selected topics / editors, Estelle Trengove, Foster Chileshe Lubasi.
pages   cm
Includes bibliographical references.
ISBN 9789351305453 (International edition)

1. Lightning protection--Africa--Congresses.   I. Trengove, Estelle, editor. II. Lubasi, Foster Chileshe, editor. III. Centre for Science and Technology of the Non-Aligned and Other Developing Countries. issuing body. IV. Title.

DDC 693.898096   23

Centre for Science and Technology of the Non-Aligned and Other Developing Countries (NAM S&T Centre)
Core-6A, 2nd Floor, India Habitat Centre, Lodhi Road,
New Delhi-110 003 (India)
Phone: +91-11-24644974, 24645134, Fax: +91-11-24644973
E-mail: namstct@gmail.com
Website: www.namstct.org

Published by   : **Daya Publishing House®**
                         A Division of
                         Astral International Pvt. Ltd.
                         – ISO 9001:2008 Certified Company –
                         4760-61/23, Ansari Road, Darya Ganj
                         New Delhi-110 002
                         Ph. 011-43549197, 23278134
                         E-mail: info@astralint.com
                         Website: www.astralint.com

Laser Typesetting   : **Classic Computer Services**, Delhi - 110 035
Printed at   : **Thomson Press India Limited**

Foreword

Lightning is a gigantic electrostatic discharge produced by unbalanced electric charges in the atmosphere either inside clouds, or cloud to cloud, or cloud to ground ad accompanied by loud thunder sound. Lightning strikes cause tremendous loses each year through fires and vulnerability of transmission towers, communication towers, and transmission lines and other tall physical structures and pose great threat to lives and property. Though most countries experience lightning strikes, some of these in the African content especially Uganda, have recorded the highest lightning densities in the world that result in huge loses of human and animal life every year. The continent's large rural population is even more vulnerable to lightning injuries than people living in cities as the people living in rural areas often work outdoors and their homes and work places are informal structures with no electrical or water reticulation that might offer a path for a lightning current to ground.

I am delighted to learn that for the capacity building in the counties of the African region in various aspects of Lightning Protection, the Centre for Science and Technology of the Non-Aligned and Other Developing Countries (NAM S&T Centre) had organized an 'African Regional Training Programme on Lightning Protection' in Kampala Uganda during 4th – 8th February 2013 jointly with the State House, Uganda and various Ministries and Agencies of the Government of the Republic of Uganda. I understand that the International Training Programme covered extensive discussions on basic characteristics of lightning, lightning safety methods, protection of equipment and structures, low cost protection schemes, first aid; counselling of victims and relatives of the dead and development of national policies on lightning safety of public which were greatly beneficial to the participants from several countries.

The present book titled 'Selected Topics in Lightning Science and Lightning Protection' has been published by the NAM S&T Centres as an outcome of these deliberations and comprises 11 scientific and status papers, which have been edited by two eminent experts on the subject, Prof. Estelle Trengove of the University of the Witwatersrand, South Africa and Mrs. Foster Chileshe Lubasi of the National Institute for Scientific and Industry Research, Lusaka, Zambia.

I am sure that this publication will serve as a useful reference material for the professionals working in the areas of Lightning Protection and will help the African countries in working out appropriate strategy, law, policies and programmes for the protection of life and property from the dangers of Lightning. I compliment the NAM S&T Centre, particularly its Director, Prof. Arun P. Kulshreshtha and his team for this initiative and bring out this very important publication for wider dissemination of the knowledge of Lightning Protection among the members of the scientific community and for enhanced South-South cooperation.

The above scientific initiative and programme is in line with Uganda Government's' policy towards enhancement of science education, research and development as a cornerstone for socio-economic advancement of society. Survival of mankind is dictated by the level of appreciation of the natural processes, particularly during the current times of what appears to be rapid global negative environmental degradation relative to current life species, as a result of irresponsible human activities and other natural factors.

HE Engineer Hilary Onek (MP)

Minister of Relief, Disaster, Preparedness and Refugees of the Republic of Uganda,

Uganda

Preface

This volume is an extended and updated version of papers presented at the African Regional Training Programme on Lightning Protection, which was held in Kampala, Uganda from 4th–8th February 2013 by the Centre for Science and Technology of the Non-Aligned and Other Developing Countries (NAM S&T Centre) in association with State House, Uganda. The Training Programme was designed to exchange experiences, including updates on available global best practices in all aspects of lightning protection.

Three of the world's leading experts on different aspects of lightning gave presentations at the workshop. Prof Vlad Rakov gave a presentation on the physics of lightning. Prof Mary Ann Cooper spoke about the different mechanisms of lightning injury. Prof Christian Bouquegneau presented an overview of the IEC 62305 International Standard on Lightning Protection.

The Training Programme was attended by delegates from several African countries, including Zimbabwe, Kenya, The Gambia, Zambia, Uganda, Togo, Tanzania, Malawi and South Africa.

An unacceptably large number of people in Africa are killed by lightning every year. The continent's large rural population is more vulnerable to lightning injuries than people living in cities. Rural people often work outdoors and their structures are often informal structures with no electrical or water reticulation that might offer a path for a lightning current to ground. Beaula Chipoyera offered some interesting perspectives in this regard. She said that if Zimbabwean women were empowered through the provision of basic facilities such as water, more accessible energy sources, and improved technology in farming such as the use of planters and harvesters, fewer women would be killed by lightning.

It became clear during the course of the workshop that there is a great need for the dissemination of information on lightning safety and lightning protection.

It was, however, encouraging to see that there are pockets of excellence in Africa, as was demonstrated by the excellent paper by Baboucarr Awe, describing how lightning damage to power lines has been reduced significantly by implementing lightning protection measures.

Although Dr. R. J. Akello did not attend the conference, it is worth noting from his paper that some commendable work is being done in Kenya as can be seen from efforts being made in lightning protection research and in sensitizing and educating the public.

An initiative that arose from the workshop, was the establishment of the African Centre for Lightning and Electromagnetics (ACLE), which will be hosted by the Makerere University Business School in Kampala. Dr. Mary Ann Cooper is the Director of the ACLE. The aim of the ACLE is to encourage lightning research in Africa. This volume of papers makes a good start towards establishing a culture of lightning research in Africa.

Estelle Trengove
Chileshe Lubasi

Introduction

Lightning has played a crucial role in the evolution of life on our planet. Lightning lights up the sky fifty to one hundred times per second all around the globe, but not every lightning bolt reaches the ground such that about 75 per cent of the lightning discharges remain in the clouds. Lightning is a spectacular natural phenomenon, but it is also a severe hazard to human life and to various objects and systems. Worldwide, there are a few thousand lightning deaths and tens of thousands of lightning injuries each year, most of which take place in developing countries. Lightning initiates many forest fires, and over 30 per cent of all electric power failures are lightning related. A lightning strike to an unprotected or inadequately protected object can be catastrophic.

Developing countries face many challenges in taking measures on lightning protection and safety. While evolving their own protection systems, it is important that the nations around the globe share knowledge, experience and technological know-how in this area so as to equip and protect themselves against this dangerous and deadly natural hazard. The annual toll of lightning casualties can be substantially reduced through public education and awareness, and serious lightning damage can be alleviated by adopting adequate lightning protection and warning systems.

With the above in view, the Centre for Science and Technology of the Non-Aligned and Other Developing Countries (NAM S&T Centre) organised an 'African Regional Training Programme on Lightning Protection' at Kampala, Uganda during 4th–8th February 2013 jointly with the Science and Technology Department of State House, Uganda and the Department of Meteorology, Uganda. The International Training Programme was intended to benefit senior managers of government departments and ministries, NGOs, scientists and scientific managers in private sector organisations with relevant background and active involvement in the subjects related to Lightning Protection. The event covered extensive discussions on basic characteristics of lightning; lightning safety methods; protection of equipments and structures; low cost protection schemes; Kerauna medicine and first aid; counselling of victims and relatives of the dead; and development of national policies on lightning safety of public. This programme was attended by 69 participants from 17 countries, including Bangladesh, Belgium, the Gambia, India, Iraq, Kenya, Malaysia, Pakistan, South Africa, Tanzania, Togo, Turkey, United Arab Emirates (UAE), United States of America (USA), Zambia, Zimbabwe and the host country Uganda.

The present publication contains 11 scientific and technical papers, of which 10 were presented during the Training Programme and one has been contributed by an invited expert on the subject. I acknowledge with gratitude the deep involvement and determined efforts of Prof. Estelle Trengove of the University of the Witwatersrand, South Africa and Mrs. Foster Chileshe Lubasi of the National Institute for Scientific and Industry Research, Lusaka, Zambia in editing this valuable publication. Last, but not the least, I appreciate the valuable services rendered by the entire team of the NAM S&T Centre, in particular, Mr. M. Bandyopadhyay for overall supervision, and Mr. Rohan Dev Talwar, Ms. Subhashree Basu and Mr. Pankaj Buttan in compiling the presented papers, liaising with the authors and editors, and giving a final shape to the volume.

I am sure that this book would serve as a useful reference material not only for the scientists and professionals engaged in the field of lightning, but also for the concerned authorities of various countries in working out appropriate strategy, laws, policies and programmes for the protection of life and property from the dangers of lightning.

Prof. Dr. Arun P. Kulshreshtha
Director and Executive Head,
NAM S&T Centre

Contents

Chapter 1

The Lightning Protection IEC 62305 International Standard

Christian Bouquegneau

University of Mons, Polytechnics, Rue de Houdain 9,
B-7000 Mons Belgium
Phone: +32 65 374040
E-mail: christian.bouquegneau@umons.ac.be

ABSTRACT

The first edition of the IEC 62305 Lightning Protection international standard was published in January 2006. This standard contains four parts [1]:

1: General principles (lightning protection, test parameters);

2: Risk management (risk assessment method, risk components);

3: Physical damage to structures and life hazard (lightning protection systems, its "Surge" only protection measures, design, installation, maintenance and inspection);

4: Electrical and electronic systems within structures (lightning Electro-Magnetic, EM pulses management of Lightning Electro-Magnetic Pulses, LEMP, earthing, bonding, magnetic shielding, line routing, Surge Protection Devices,SPD, systems).

This first edition was broadly accepted around the world. It states that a global approach is needed to address the phenomenon in a correct and comprehensive way. All parameters are interconnected through the four parts, especially to enter part 4 related to protection against LEMP: surge protective devices, shielding, cable routing. Moreover, IEC 61643 standards (from IEC SC37A) are in complete conformity with IEC 62305 (from IEC TC 81).

In the second edition, published in 2010, the text was improved for clarity sake and some equations and collection areas were simplified. Explosive areas (explosive zones 1, 2 and 21, 22) are now normative (instead of informative).

Part 2 of the IEC 62305 standard (Risk Management, IEC 62305-2) addresses the risk management aspect of lightning protection. Lightning hazards can cause damage to structures, the contents in the structures, failure of associated electrical and electronic systems and injury to living beings in or near the structures. Consequential effects of the damages and failures may be extended to the surroundings of the structure or involve the environment. The risk, defined in this standard, is the probable mean annual loss in a structure due to lightning flashes. The lightning risk depends on a lot of parameters (about 70). All four sources of damages (direct strike to the structure, strike to ground near the structure, direct strike to the incoming lines, strike to ground near the incoming lines) need to be addressed to calculate the risk.

The author is a member of the maintenance groups and convenor of some working groups and task forces now preparing the third edition. After looking at the main points of the existing standard, he will introduce some new trends under consideration.

Keywords: Lightning, Lightning protection, Risk assessment, International standards.

1. General Principles of Effective Lightning Protection

Sometimes it is difficult and even costly, but nowadays it is possible by applying some prevention rules and standards, namely the last ones issued by the Technical Committee 81 (TC 81: Lightning Protection) in the International Electrotechnical Commission (IEC). In Europe, the Technical Committee TC 81X in CENELEC follows the same rules. These standards (IEC 62305) first issued in 2006, then in 2010 [1], for the second edition, are based on the concept of *risk management* in order to significantly reduce damages due to lightning on protected structures.

Indeed, so far, there exists neither devices nor methods capable of preventing lightning discharges. Direct *cloud-to-ground* flashes or nearby structures are hazardous to people and to structures, their contents and installations, as well as to services. This is why lightning protection measures have to be applied, taking into account risk evaluation and management.

Concerning the external lightning protection and loss of human life, very little progress has been made since Franklin *lightning rods* (1752) and the development of *meshed cages* (Faraday cages surrounding entirely the building or the structure to be protected). The positioning of an external lightning protection system (ELPS), a better name than lightning rod which makes people think of a single vertical rod, has to be carefully studied for the design of a new structure in order to take into consideration the use of conductive elements of the structure for earthing. Particular care must be taken in order to minimize the electric resistance.

Generally, an external lightning protection system is not sufficient. Services (power lines, telecommunication lines.) entering the building or structure and the protection against lightning electro-magnetic pulses (LEMP) must be taken into account. Surge protective devices (SPDs) and magnetic shielding have constantly improved. Thanks to better knowledge of lightning current characteristics. SPDs presenting a large resistance to lightning currents are installed in parallel with an electrical circuit; they do not act when applied voltage is under the threshold voltage and are perfectly conductive above the threshold level and lightning over-voltages.

1.1 The Electro-geometric Model (EGM)

A downward *stepped leader* initiated from the *negative* charge part of the cumulonimbus wearing a negative charge concentrated at its tip proceeds to ground and increases the vertical component of the electric field to values around 300 kV/m. Positive partial discharges of the corona type propagate to the downward leader tip. The one which first meets the downward leader tip leads to a return stroke jumping from ground through the completely ionized channel.

The vertical component of the electric field nearby ground depends on the charge amount existing in the leader and on the distance separating the tip from the ground. The first stroke generally neutralizes the leader charge so that it exists in form of a quasi constant ratio between the electric charge (C) and the current peak (kA). Empirically, this ratio is about15 kA/C.

From this hypothesis, a relation was established linking the *striking distance* d and the presumed current amplitude I, when the critical electric field is reached at ground level.

The electro-geometric model takes this dependence into account though it does not discriminate among various types of the attracting structure (ground, tree, building, lightning rod.), nor among different heights. Strictly speaking this model is only applicable to flashes with negative downward leaders (90 per cent of such flashes in temperate regions). Though it is not completely proven scientifically, the electrogeometric model now looks like the best model so far since it is entirely validated statistically worldwide. Various new models based on the electrogeometric one take into account preferential attractive effects at larger heights but they are not confirmed yet.

According to the electro-geometric model, the striking distance d (meters) is related to the current amplitude (kA) by the relation (Figure 1.1):

$$d = 10 \, I^{0.65} \tag{1.1}$$

The object which is closer to the leader tip at a distance d from its tip constitutes the strike flash point on condition that this object is earthed. The interception leader progresses more rapidly when the ground impedance of the object is lower.

Though the electro-geometric model was only developed for the negative downward leaders, it is also used for positive leaders the tips of which initiate upward streamers from a corona effect. However, when such a leader approaches the ground, interception leaders rarely develop. It is considered that corona streamers play the same role as in the case of negative discharges.

As an example, let us consider Figure 1.2 which illustrates the protection of a vertical rod of height h (h = 80 m) for two different values of the striking distance d_1 (15 m) and d_2 (100 m) corresponding to current amplitudes I equal to 1.9 kA and 35 kA respectively. When the striking distance d is small (see figure 2a where d_1 = 15 m), the rod only protects a small curvilinear cone at its bottom and lightning can reach almost any part of the rod or the ground farther than distance OA_1 = 15 m; on the other hand, if the striking distance is larger than the rod height (see figure 2b where d_2 =

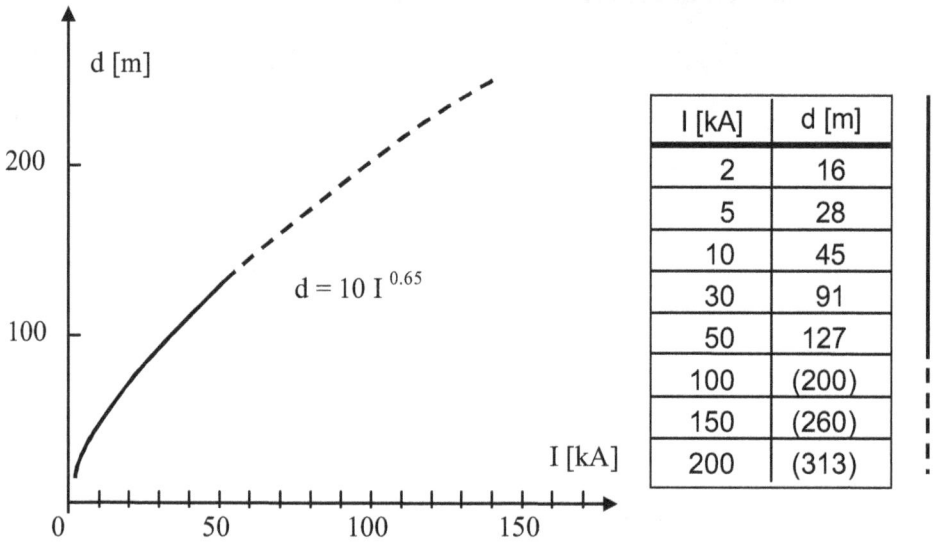

Figure 1.1: Striking Distance According to the Electro-geometric Model.

I [kA]	d [m]
2	16
5	28
10	45
30	91
50	127
100	(200)
150	(260)
200	(313)

$d = 10\,I^{0.65}$

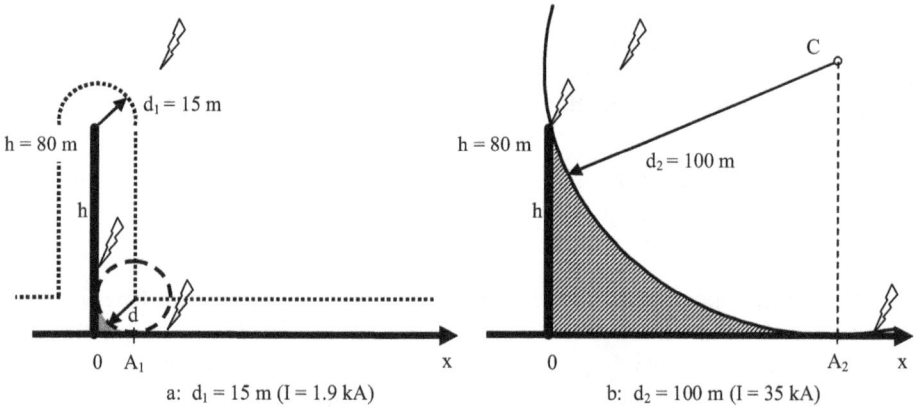

a: $d_1 = 15$ m (I = 1.9 kA)

b: $d_2 = 100$ m (I = 35 kA)

Figure 1.2: Electro-geometric Model Applied to the Protection of a Vertical Rod.

100 m), any object situated inside the big curvilinear cone will be protected and the lightning current will reach either the rod tip or the ground at a distance larger than OA_2. Note that the protection zone is larger when the current amplitude is larger.

When the current amplitude is larger than 50 kA, the dispersion is larger and larger and the strike is more and more hazardous. On the contrary, at small current amplitudes (order of one kA), the protection zone of the vertical rod is reduced. It is also impossible to get a perfect protection by means of a single vertical rod installed on a building. A complete Faraday cage or a net of suspended wires are usually used. Particular attention is paid to the electrical continuity between wires and between wires and ground.

1.2 Lightning Protection of Buildings and other Structures

So far it has been impossible to prevent lightning from striking, as there is no device or method to avoid lightning flashes. Cloud-to-ground discharges to or nearby structures can harm these structures, their contents and persons. Lightning protection measures must be applied in terms of risk management.

Risk analysis helps with the determination of the required level of protection. Choosing the right protection level is crucial. The objective is to reduce the risk of degradations due to a direct lightning flash on the structure or inside the volume to be protected under a specific value called *maximum tolerable risk*. Damages depend on several things such as the use, the content (persons and things) to be protected, the materials used and the protection measures imposed to reduce the lightning risk.

The lightning current characteristics are chosen from world data collected by CIGRE (International Council for High-Voltage Transmission Lines). Wave-shapes are relevant to several categories: components of short or large durations, possible components of downward flashes depending on polarity, components of upward flashes depending not only on polarity but also on the possible succession of different components. The wave-shapes lead to specific design of lightning protection systems against electromagnetic pulses generated by lightning as well as tests on lightning protection devices.

In the international standard IEC 62305 ([1]), four protection systems are defined (I, II, III, IV, respectively). Each one corresponds to a set of construction rules and is bound to a related protection level (I, II, III or IV). As a first step, a *global protection efficiency* was associated to 98 per cent (level I), 95 per cent (level II), 90 per cent (level III) or 80 per cent (level IV). To obtain a protection level higher than 98 per cent (level I+), additional lightning protection measures have to be applied. A set of *minimum and maximum values* of the lightning current is associated to each protection level (I, II, III or IV).

Maximum values of the lightning peak current for the first short stroke (short strokes have a duration smaller than 2 ms, long strokes have a duration larger than 2 ms) are 200 kA (99 per cent of flashes) for level I, 150 kA (98 per cent of flashes) for level II and 100 kA (97 per cent of flashes) for levels III and IV. For the current capability design of SPDs (lightning current surge protective devices), it is assumed that 50 per cent of this current flows into the external lightning protection system (earth-termination system included) and 50 per cent through the services within the structure to be protected. Should the service consist solely of a three-phase power supply (3 phases + neutral = 4 lines) then the following design currents could be expected: 25 kA for level I, 18,75 kA for level II and 12.5 kA for levels III and IV. In reality, multiple connections are used dividing the currents further and reducing their effective peak values.

Minimum values of the lightning current peaks are linked to the application of the rolling sphere method (the rolling sphere radius d is equal to the striking distance) in the design of lightning protection systems. They are selected as 3 kA (99 per cent of flashes; d = 20 m) for level I, 5 kA (97 per cent of flashes; d = 30 m) for level II, 10 kA (91 per cent of flashes; d = 45 m) for level III and 16 kA (84 per cent of flashes; d = 60 m) for level IV. A weighted probability is easily imposed because current parameters lie between minimum and maximum values and a set of protection measures is defined for this range of values by the protection level selected. The efficiency of such protection measures is supposed to be equal to the probability of lightning current parameters remaining inside the range.

An external lightning protection system assures the more efficient protection of persons, buildings and other structures. The type and the positioning of the conductive elements of the installation fit the efficiency concerns as well as the aesthetic ones and try to minimize the costs. Special attention is paid to the design of an earth termination system with a resistance as small as possible.

The external lightning protection system (tri-dimensional Faraday cage), consists of:

☆ an air-termination system with attracting elements (horizontally meshed metallic conductors, vertical rods for non-conductive masts, suspended wires from tower to tower), avoiding unuseful « aigrettes » (on the contrary, they lead to a more uniform electric field in the ambient air and tend to reduce the point effect), radioactive rods (now forbidden in most of the countries) and so-called active rods (of any type till now);

☆ a down-conductor system (metallic conductors generally vertically binding with the air-termination system to measuring connections);

☆ Ring conductors or *equipotential bonding* (for elevated buildings) (metallic conductors forming a horizontal loop around the structure assuring the electric equipotentialization with the down-conductor system);

☆ Measuring connections (connection which can be dismantled between a down-conductor and an earth conductor allowing the measuring of the ground conductor resistance);

☆ Earth-termination system (buried metallic conductors ensuring electric continuity with ground), eventually connected to the ring earth electrodes (foundation earth electrode or binding loop, binding two or several earth electrodes generally surrounding the structure to be protected).

The electrical continuity between different metallic parts is achieved by soldering or brazing. Bound structures are made equipotential.

Some natural components are considered to be parts of the air-termination system when they conform with the requirements imposed to the elements artificially introduced (electrical continuity, sufficient thickness and cross-section.).

Thanks to the application of the electro-geometric model, the air-termination system is positioned in relation with the protection level selected (a rolling sphere radius equal to d = 20, 30, 45 or 60 m is achieved for protection levels I, II, III or IV respectively).

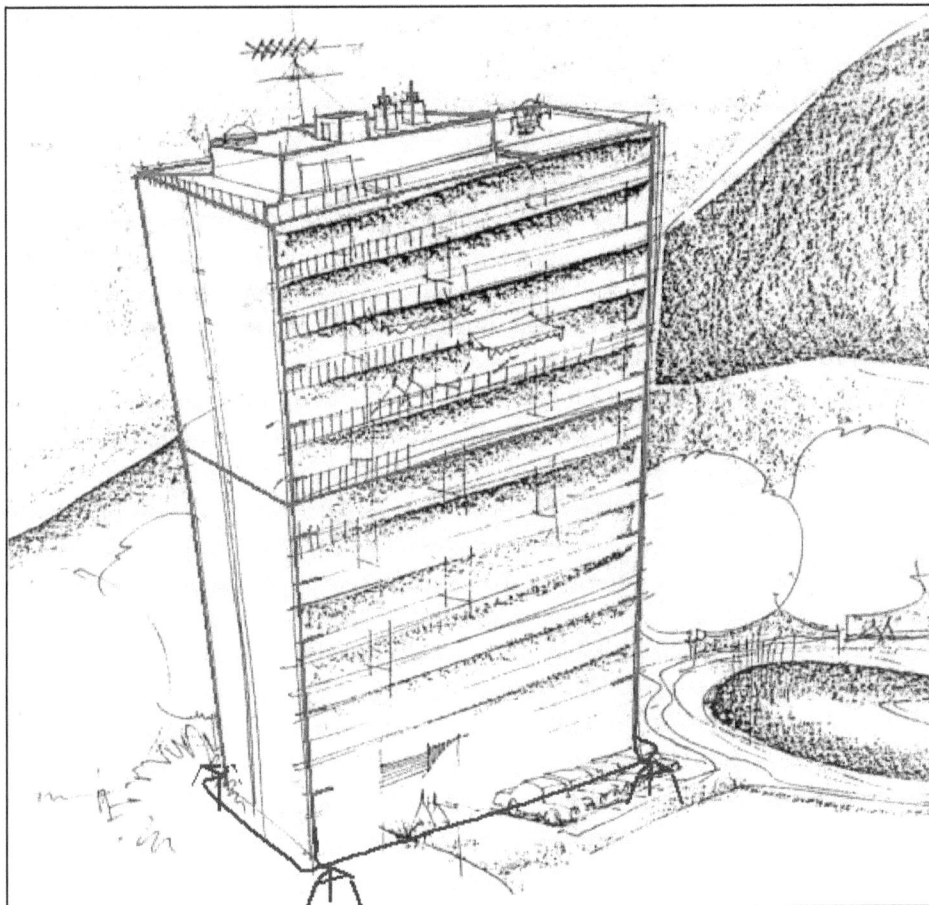

Figure 1.3: Ideal Lightning Protection of a Building.

On the roof, this type of protection is similar to a flat conducting mesh of sides 5 m (level I), 10 m (level II), 15 m (level III) or 25 m (level IV).

The positioning of the exterior lightning protection system must be carefully studied from the early design of a new structure to make use of the metallic parts of the structure, to obtain as low an earth resistance as possible.

In buildings higher than 20 m, down-conductors are interconnected with horizontal ring conductors about 20 m apart.

Down-conductors are distributed along the structure perimeter so that their mean inter-distance does not exceed 10 m, 15 m, 20 m or 25 m for protection levels I, II, III or IV respectively. In all cases, a minimum of two down-conductors is required.

Properly shaping and dimensioning earth electrodes are more important than obtaining a desired minimum value of the earth resistance. Nevertheless, when possible, the highest value of earth resistance should not exceed 10 ohms! C.

Bouquegneau and B. Jacquet showed ([5]) that little deep electrodes (vertical or inclined) interconnected by a ring earthing conductor (foundation earth electrode) would be advantageous to use (Figure 1.3).

2. Lightning Risk Management

2.1 Lightning Damage and Lightning Risk

Hazards to a structure can result in damage to the structure and to its contents, failure of associated electrical and electronic systems, injury to living beings in or close to the structure and damage to a service utilities like power lines; telecommunication networks and data lines; water, gas and fuel networks entering the structure.

Part 2 of the new standard IEC 62305 ([1.2]) addresses the risk management of protection against lightning. In this second part, damage to a structure is subdivided into four sources of damage (S1, S2, S3, S4) and three types of damage (D1, D2, D3). The lightning current is the primary source of damage; the following sources are distinguished by the strike attachment point; three basic types of damage can occur as a result of lightning flashes.

☆ Sources of damage:

S_1 : Flashes to a structure;

S_2 : Flashes near a structure;

S_3 : Flashes to a service utility;

S_4 : Flashes near a service utility.

☆ Types of damage:

D_1 : Injury to living beings (due to step and touch voltages)

D_2 : Physical damage (fire, explosion, mechanical destruction, chemical release. due to lightning current effects including sparking)

D_3 : Failure of electrical and electronics systems (due to lightning electromagnetic pulses, LEMP).

Flashes to the structure (S_1) may cause physical damage and life hazards. Flashes near the structure (S_2) or service (S_4) as well as flashes to the structure (S_1) or service (S_2) may cause failure of electrical and electronic systems due to over-voltages resulting from resistive and inductive coupling of lightning current to these systems.

The number of lightning flashes affecting the structure and the services depends on the dimensions, the characteristics of the structure and of the services. It also depends on the characteristics of the environment around the structure and the services, as well as the lightning ground flash density in the region where the structure and the services are located. The probability of lightning damage depends on the structure, the services, the current characteristics, and the type and efficiency of applied protection measures. The annual mean amount of the consequential loss depends on the extent of damage and the consequential effects that may occur as result of a lightning flash.

Point of strike	Example	Source of damage	Cause of damage	Type of damage
Structure		S_1	C_1 C_2 C_3	$D_1, D_4{}^b$ D_1, D_2, D_3, D_4 $D_1{}^a, D_2, D_4$
Earth near the structure		S_2	C_3	$D_1{}^a, D_2, D_4$
Entering supply line	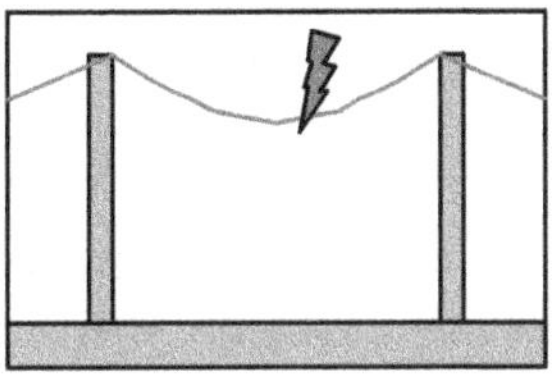	S_3	C_1 C_2 C_3	$D_1,$ D_1, D_2, D_3, D_4 $D_1{}^a, D_2, D_4$
Earth near entering supply line	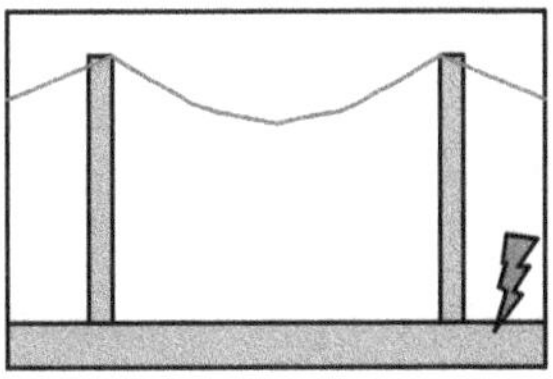	S_4	C_3	$D_1{}^a, D_2, D_4$

[a] For hospitals and structures with risk of explosion or other structures where failures of internal systems endangers human life.   [b] For agricultural properties (loss of animals).

Figure 1.4: Sources and Types of Damage.

Each type of damage, alone or in combination with others, may produce a different consequential loss in the object to be protected. The type of loss that may occur depends on the characteristics of the object and its content. The following types of loss shall be taken into account (Figure 1.4): L_1: loss of human life; L_2: loss of service to the public; L_3: loss of cultural heritage; L_4: loss of economic value (structure and its content, service, loss of activity). All four may be associated with a structure.

The primary risks to be evaluated in a structure correspond to their equivalent type of loss and may then be R_n (n = 1, 2, 3, 4):

R_1: Risk of loss of human life (by far the most important risk to consider!);

R_2: Risk of loss of service to the public (implications of the public loosing its gas, water or power supply or, more generally, risk related to the provider which cannot provide its service to its customers);

R_3: Risk of loss of cultural heritage (all historic buildings and monuments, with focus on the loss of the structure itself);

R_4: Risk of loss of economic value (which compares, amongst other factors, the cost of the loss in an unprotected structure to that with protection measures applied).

Protection against lightning is required if the primary risk R is greater than a tolerable risk R_T.

2.2 Identification of Lightning Risk Components

To evaluate the each primary risk R_n (n = 1, 2, 3, 4), the relevant risk components R_x (partial risks depending on the source S and the type of damage D) shall be defined and calculated. Each primary risk R_n is the sum of its risk components R_x. When calculating a risk, the risk components R_x may be grouped according to the source of damage (S) and the type of damage (D):

* risk components for a structure due to flashes to the structure (S_1):

R_A : Component related to injury to living beings caused by touch and step voltages in the zones up to 3 m outside the structure; L_1 and, in the case of structures holding livestock, L_4 because of the possibility of losing animals;

R_B : Component related to physical damage caused by dangerous sparking inside the structure triggering fire or explosion, which may also endanger the environment : L_1, L_2, L_3 and L_4 may arise;

R_C : Component related to failure of internal systems caused by LEMP : L_2 and L_4 could occur. L_1 could occur in the case of structures with risk of explosion and hospitals or other structures where failure of internal systems immediately endangers human life;

* risk component for a structure due to flashes near the structure (S_2):

R_M : Component related to failure of internal systems caused by LEMP : L_2 and L_4 could occur as well as L_1 in the case of structures with risk of explosion and hospitals or other structures where failure of internal systems immediately endangers human life;

* risk components for a structure due to flashes to a service connected to the structure:

R_U : Component related to injury to living beings caused by touch voltage inside the structure, due to lightning current injected in a line entering the structure; L_1 and, in the case of agricultural properties, L_4 with possible loss of animals may also occur;

R_V : Component related to physical damage (fire or explosion triggered by dangerous sparking between external installation and metallic parts generally at the entrance point of the line into the structure), due to lightning current transmitted through or along incoming services; L_1, L_2, L_3 and L_4 may occur;

R_W : Component related to failure of internal systems caused by over-voltages induced on incoming lines and transmitted to the structure : L_2 and L_4 could occur along with L_1 in the case of structures with risk of explosion and hospitals or other structures where failure of internal systems immediately endangers human life;

* risk component for a structure due to flashes near a service connected to the structure:

R_Z : Component related to failure of internal systems caused by over-voltages induced on incoming lines and transmitted to the structure : L_2 and L_4 could occur along with L_1 in the case of structures with risk of explosion and hospitals or other structures where failure of internal systems immediately endangers human life.

Risk components R_x to be considered for each type of loss (L_1 to L_4) in a structure are listed below :

$$R_1 = R_A + R_B + R_C^{(1)} + R_M^{(1)} + R_U + R_V + R_W^{(1)} + R_Z^{(1)}; \tag{2.1}$$

$$R_2 = R_B + R_C + R_M + R_V + R_W + R_Z; \tag{2.2}$$

$$R_3 = R_B + R_V; \tag{2.3}$$

$$R_4 = R_A^{(2)} + R_B + R_C + R_M + R_U^{(2)} + R_W + R_Z. \tag{2.4}$$

(1) Only for hospitals and structures with risk of explosion or other structures when failure of internal systems endanger human life. (2) Only for properties where animals may be lost.

These risk components, which can be composed with reference to the source of damage (S_1 to S_4) or the type of damage (D_1 to D_3) are better defined by looking at Figures 1.5 and 1.6.

Each primary risk R_n can be expressed with reference to the source of damage and can be split into two (direct R_D related to S_1 and indirect R_I related to S_2, S_3 and S_4) basic components (for each loss):

$$R_n = R_D + R_I \tag{2.5}$$

with

$$R_D = R_A^{(2)} + R_B + R_C^{(1)} \tag{2.6}$$

and

$$R_I = R_M^{(1)} + R_U + R_V + R_W^{(1)} + R_Z^{(1)} \tag{2.7}$$

(1) Only for hospitals and structures with risk of explosion or other structures when failure of internal systems endanger human life. (2) Only for properties where animals may be lost.

Each risk component can be evaluated by the equation

$$R_X = N_X P_X L_X \tag{2.8}$$

where N_X is the number of dangerous events per annum, P_X is the probability of damage to a structure and L_X is the amount of loss to a structure.

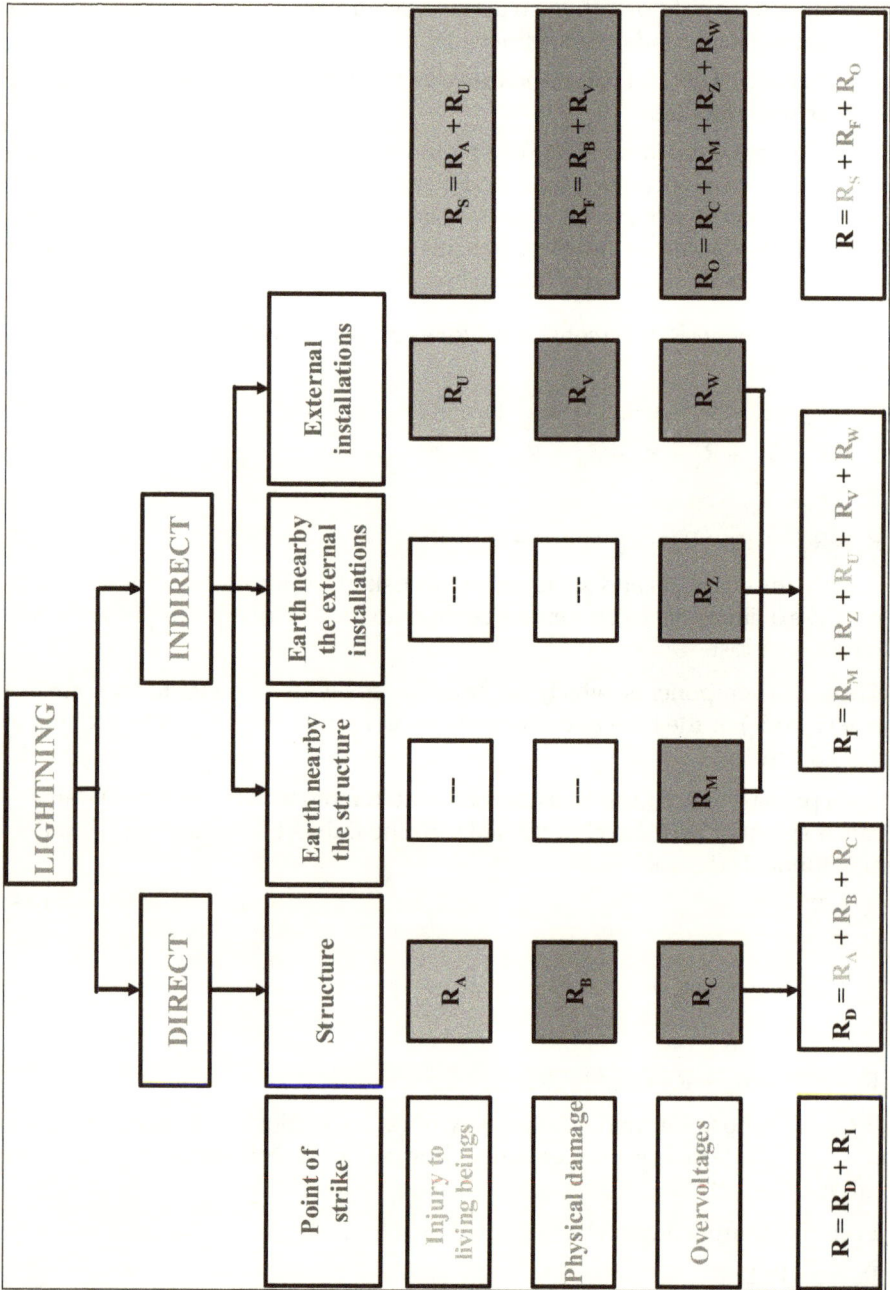

Figure 1.5: Composition of the different Risk Components.

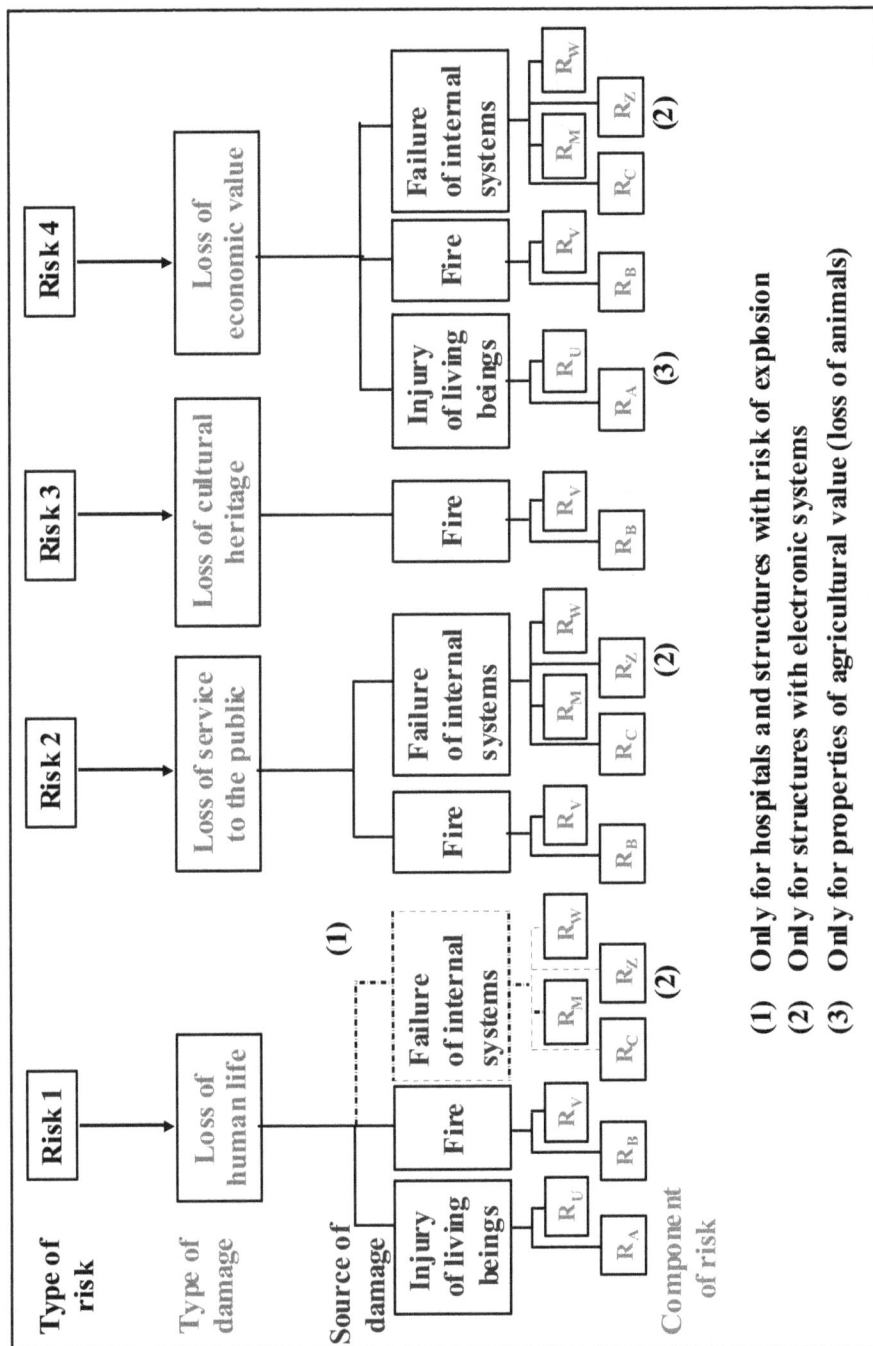

Figure 1.6: Types of Loss Resulting from different Types of Damage.

(1) Only for hospitals and structures with risk of explosion
(2) Only for structures with electronic systems
(3) Only for properties of agricultural value (loss of animals)

The values of N_X (number of dangerous events per annum) can be evaluated (see annex A in [1.2]). For example,

$$N_D = N_g A_{d/b} C_{d/b} 10^{-6} \tag{2.9}$$

where N_g is the lightning ground flash density (km^{-2}year^{-1}), $A_{d/b}$ is the collection area of the isolated structure (m^2) and $C_{d/b}$ is the location factor of the structure; for an isolated rectangular structure with length L, width W and height H on a flat ground, the collection area is equal to

$$A_{d/b} = L W + 6 H (L + W) + 9 \pi H^2 \tag{2.10}$$

the location factor $C_{d/b}$ is equal to 0.25 if the object is surrounded by higher objects or trees, 0.5 if t is surrounded by objects or trees of the same heights or smaller, 1 if it is isolated and 2 if it is isolated on a hilltop or a knoll.

The probability of damage to a structure P_X is assessed by applying concepts described in the informative annex B in [1.2]. The amount of loss L_X is assessed by applying concepts described in the informative annex C in [1.2]).

The decision to protect a structure against lightning, as well as the selection of protection measures, shall be performed following the flow diagram shown in Figure 1.7.

The designer initially identifies the types of loss that could result from damage due to lightning. He determines the primary risk R of each type of loss identified and identifies the corresponding tolerable risk R_T. It is the responsibility of the authority having jurisdiction to identify the value of tolerable risk R_T. Representative values of the acceptable risk R_a or tolerable risk R_T are given in Figure 1.8.

The calculated risk R is then compared to the corresponding tolerable risk R_T.

If $R < R_T$, the structure is adequately protected for the particular type of loss considered.

If $R > R_T$, protection measures need to be applied to this type of loss.

A series of trial and error calculations is performed in order to reduce the risk R under R_T.

Generally, if the risk evaluation requires a structural lightning protection system (LPS), *i.e.* if $R_D > R_T$, an equipotential bonding or lightning current type I SPDs are always required for both energy lines and telecommunication lines. These SPDs constitute an integral part of the structural LPS and the first part of a coordinated set of SPDs for the protection of electronic equipment.

If $R_D < R_T$, but $R_I > R_T$, any electrical services entering the structure via an overhead line will require lightning current type I SPDs of level 12.5 kA. For underground electrical services entering the structure, protection is achieved with type II SPDs. Underground electrical services are not subject to direct lightning strikes; therefore, partial lightning currents are not transmitted to the structure so that underground electrical services do not have requirements for lightning current type I SPDs where no lightning protection system is present.

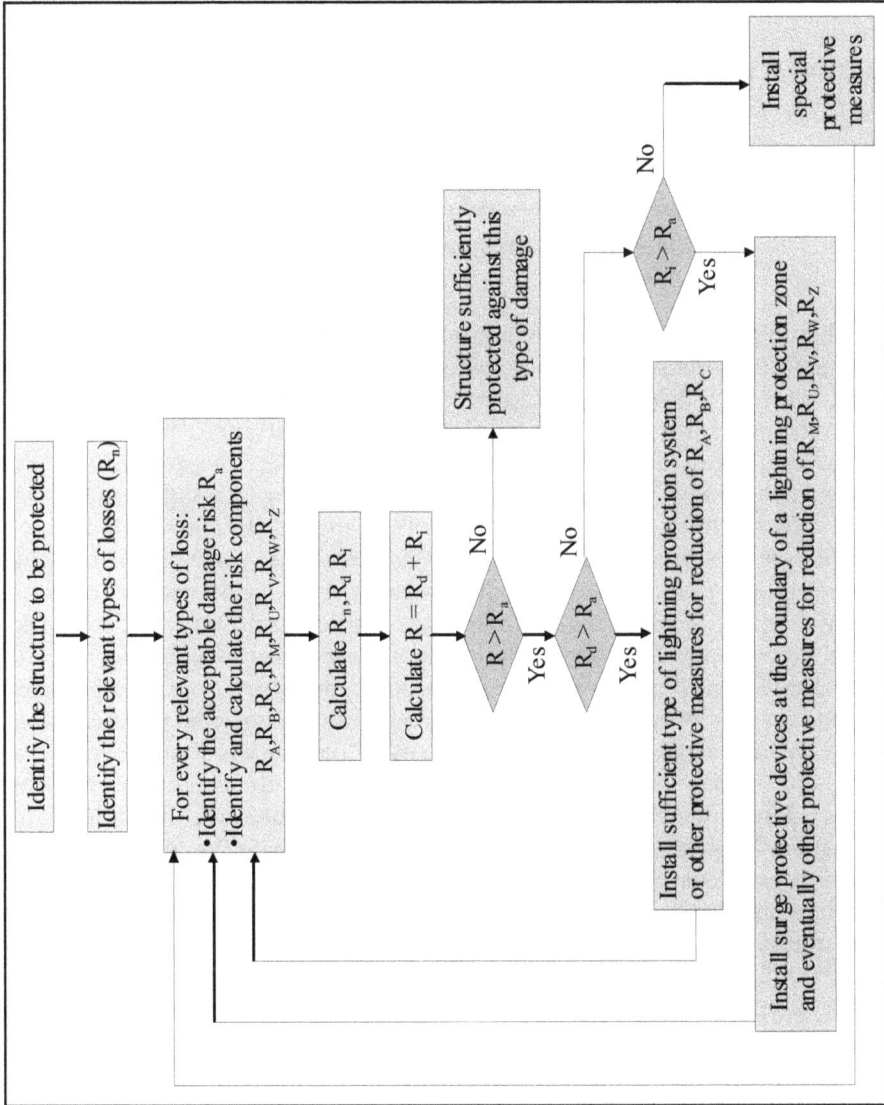

Figure 1.7: Flow Diagram Showing the Procedure for Deciding the Need of Protection.

Type of damage	R_T
Loss of human life	10^{-5}
Loss of service to the public	10^{-3}
Loss of cultural heritage	10^{-4}

Figure 1.8: Typical Values of the Tolerable Risk R_T.

The lightning risk assessment for a structure, based on the IEC 62305-2 standard, leads to numerous calculations (68 parameters in each zone selected). Useful software had to be created. We designed a complete software [7], called "RISK Multilingual 4" (to get an offer for the last updated 2013 software, just write to christian.bouquegneau@umons.ac.be); it is not only the most complete and comprehensive we know, but it is available in any language. Designed in English, French and Dutch, it can be translated in any other language by simply modifying a specific file containing a given word list. It also includes various maps of ground flash densities in different countries; other maps can easily be integrated according to National needs. "RISK Multilingual 4" takes into account all 68 parameters described in IEC 62305-2, edition 2, so as to rigorously assess the lightning risk on a structure. The user can choose any parameter values even different from those proposed in the standard. For example, entering lines length can be chosen as the user wants. The interface has been designed to clearly show all parameters simultaneously, as well as the principal results of risk assessment, namely the collection areas, the number of dangerous events experienced by the structure and the various risk components with their compositions. The user can estimate the influence of any parameter on various risk components. Thus, he can easily deduce the kind of protection required to guarantee a certain value of tolerable risk. Nevertheless, due to the large number of parameters to be considered, special help is proposed to the user thanks to tool tips provided to define each parameter considered. The user can also introduce any parameter value through windows that describe each of them. Furthermore, to evaluate the need of installing lightning protection measures, results are exported to an Excel spreadsheet. You can easily sum up all risk components in each zone and compare them to the tolerable risk imposed. The Excel file is used to write the complete report on the risk analysis of the given structure. This software was designed to be used in any language. In the same way, any new national map of ground flash densities can easily be introduced in the software.

3. External Lightning Protection System

As stated above, an external lightning protection system (external LPS or structural lightning protection system) is intended to intercept direct lightning flashes to a structure, to conduct the lightning current safely to earth and to disperse the lightning current into the earth. These three goals can be considered separately by means of three complementary systems: an air-termination system (section 3.1), a down-conductor system (section 3.2) and an earth-termination system (section 3.3).

The internal LPS (section 4) is supposed to prevent dangerous sparking within the structure to be protected, by using either equipotential bonding or an electrical insulation thanks to a separation distance (distance between two conductive parts at which no dangerous sparking can occur) between the external LPS components and other electrically conducting elements internal to the structure.

The complete LPS has to be an effective measure not only for the protection of structures against physical damage, but also for the protection against injury to living beings due to touch and step voltages (section 3.4). The main purpose is to reduce the dangerous current flowing through bodies by insulating exposed conductive parts and by increasing the surface soil resistivity. Soil conductivity and the nature of the earth are very important.

The type and location of an external LPS should primarily take into account the presence of electrically conductive parts of the structure. The external LPS can be isolated or not isolated from the structure to be protected. In an isolated external LPS, the air-termination system and the down-conductor system are positioned in such a way that the path of the lightning current has no contact with the structure to be protected and that dangerous sparks between the LPS and the structure are avoided. In most cases, the external LPS is attached to the structure to be protected.

An isolated external LPS is required when the thermal and explosive effects at the point of strike on the conductors carrying the lightning current may cause damage to the structure or to its content. Among the typical cases are structures with combustible walls and coverings as well as the areas with danger of explosion and fire.

Another important requirement is to avoid dangerous sparking between the lightning protection system and the structure; this is satisfied in an isolated external LPS by sufficient insulation or separation and in a non-isolated external LPS by bonding or by sufficient insulation or separation.

3.1 Air-termination System

The air-termination system is the first part of an external LPS using metallic elements such as rods, mesh conductors or catenary wires which is intended to intercept lightning flashes.

Air-termination systems can be composed of any combination of vertical rods (including free-standing masts), catenary wires, horizontal or meshed conductors. Individual air-termination rods are connected together at roof level to ensure the current division.

Air-termination components installed on a structure are preferably located at corners, exposed points and edges. The components are located especially on the upper level of any facades, in accordance with the rolling sphere method.

In the international standard IEC 62305, four classes (respectively I, II, III, IV) of lightning protection systems corresponding to a set of construction rules and related to four protection levels (respectively I, II, III, IV).

3.2 Down-Conductor System

The down-conductor system is the second part of an external LPS that is intended to conduct the lightning current from the air-termination system to the earth-termination system. In order to reduce the probability of damage due to lightning current flowing in the lightning protection system, several equally spaced down-conductors of minimum length are installed and an equipotential bonding to conducting parts of the structure is performed.

The choice of the number and position of down-conductors should take into account the fact that, if the lightning current is shared in several down-conductors, the risk of side flashes and electromagnetic disturbances inside the structure is reduced. It follows that, as far as possible, the down-conductors should be uniformly placed along the perimeter of the structure and with a symmetrical configuration. The current sharing is improved not only by increasing the number of down-conductors but also by equipotential interconnecting rings.

Down-conductors should be placed as far as possible away from internal circuits and metallic parts in order to avoid the need for equipotential bonding with the lightning protection system.

An equal spacing between down-conductors should never exceed 20 m, or even 10 m at level I (see table of Figure 1.9). Lateral connection of down-conductors is made not only both at the top of the structure and at ground level but also every 10 m to 20 m of height of the structure. A tolerance of about 20 per cent is generally accepted.

Class of LPS	Typical distances *(m)*
I	10
II	10
III	15
IV	20

Figure 1.9: Typical Values of Spacing between Down-Conductors and between Ring Conductors According to the Class of LPS.

In the *isolated* case, if the air-termination system consists of rods on separate non-metallic masts, at least one down-conductor is needed for each mast. If it consists of catenary wires, at least one down-conductor is needed at each supporting structure. If it consists of a network of conductors, at least one down-conductor is needed at each supporting wire end.

For each *non-isolated* lightning protection system, the number of down-conductors shall be not less than two and the down-conductors should be distributed around the perimeter of the structure to be protected.

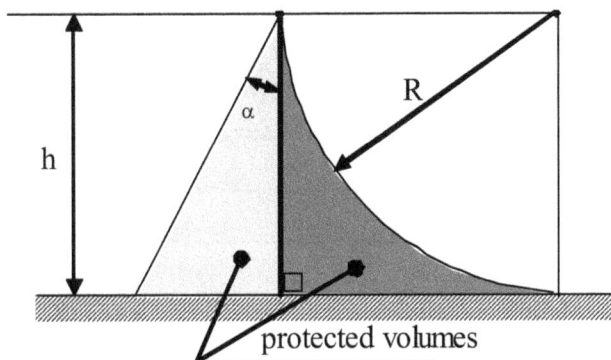

protected volumes

The table below gives the maximum values of rolling sphere radius R, mesh size W and protection angle α corresponding to the class of lightning protection systems.

| Class of LPS | Protection Method | | |
	Rolling Sphere Radius R(m)	Mesh Size W (m)	Protection Angle
I	20	5 x 5	See Figure 1.9
II	30	10 x 10	
III	45	15 x 15	
IV	60	20 x 20	

Note that the protection angle method is not applicable beyond the values marked with •; only rolling sphere and mesh methods apply in these cases; h is the height of air–termination above the area to be protected; the angle will not change for values of h below 2 m.

Equipotentialization is achieved by interconnecting the lightning protection system with the external conductive parts and lines connected to the structure, but also structural metal parts, metal installations and internal systems. Interconnecting means electrical continuity by bonding conductors (if not provided by natural bonding) or installation of surge protective devices when direct connections with bonding conductors is not feasible.

In the case of an *isolated* external lightning protection system, lightning equipotential bonding shall be established at ground level only. For a *non-isolated* external lightning protection system, bonding with connections as direct and straight as possible is achieved not only at the ground level but also every 10 m (LPS class I) to 20 m (LPS class IV) from the bottom and along the height of the down-conductors.

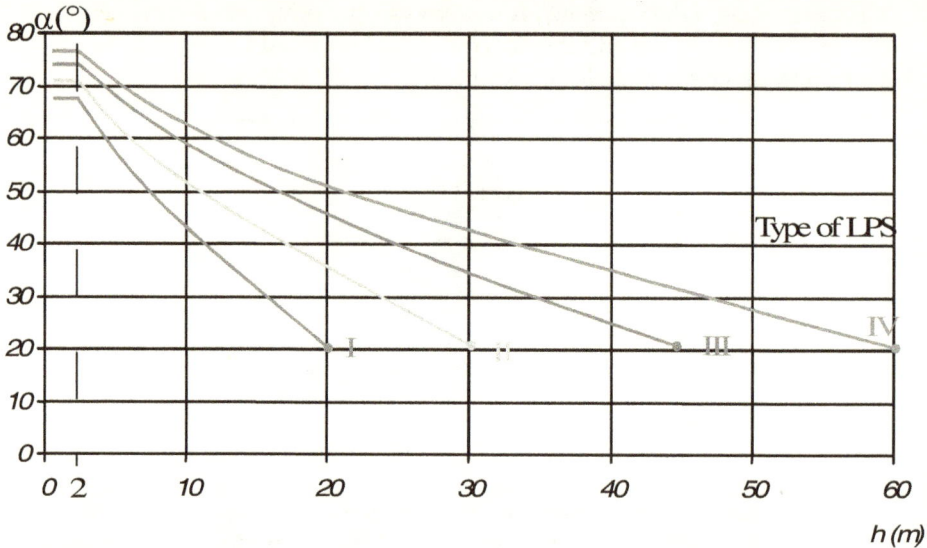

Figure 1.10: Comparison of the Three Usual Lightning Protection Methods.

3.3 Earth-termination System

The earth-termination system is the third part of an external LPS which is intended to conduct and disperse the lightning current into the earth, without causing any danger to people or damage to installations inside the structure to be protected.

The *surge impedance* is the ratio of the instantaneous value of the earth-termination voltage (potential difference between the earth-termination system and the remote earth) over the instantaneous value of the earth-termination current which in general do not occur simultaneously. The *conventional earth resistance* is the ratio of the peak values of the earth-termination voltage and the earth-termination current. The *impulse factor* is the ratio of the conventional earth resistance over the low-frequency resistance of the earth electrode.

From the viewpoint of lightning protection of single structures, a single integrated earth-termination system is preferable and is suitable for all purposes (*i.e.* lightning protection, power systems and telecommunications systems).

According to IEC 62305-3 (see [1.3]), two basic types of earth electrodes arrangements can apply:

☆ Type A arrangement comprising horizontal or vertical earth electrodes installed, outside the structure to be protected, connected to each down-conductor, with a minimum of two earth electrodes;

☆ Type B arrangement comprising either a ring conductor external to the structure to be protected, in contact with the soil for at least 80 per cent of its total length, or a foundation earth electrode.

l_1 (m)

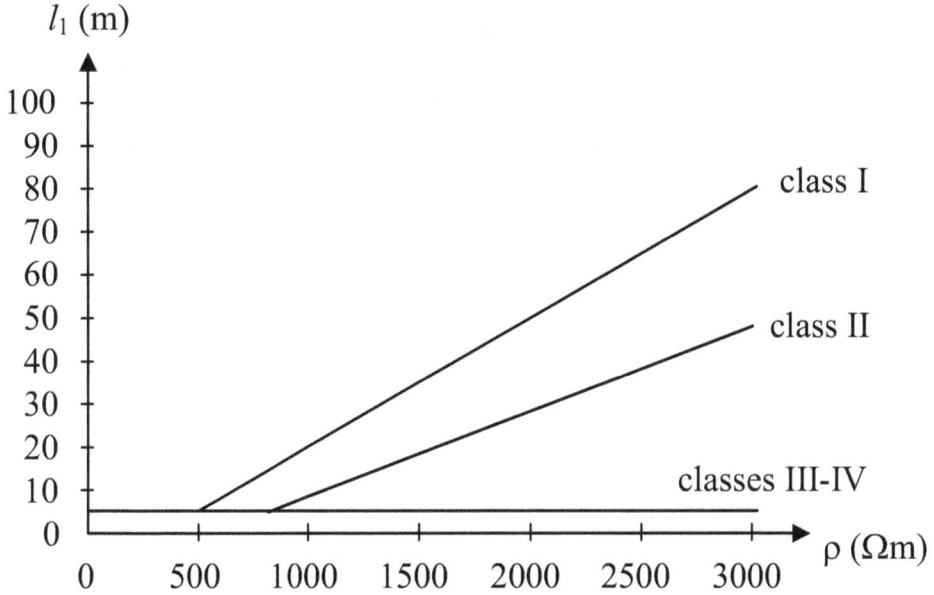

Figure 1.11: Minimum Length of an Earth Electrode According to the Class of LPS.

In a type A arrangement, the minimum length of each earth electrode at the base of each down-conductor is l_1 for horizontal electrodes or 0.5 l_1 for vertical (or inclined) electrodes, l_1 being the minimum length of horizontal electrodes shown in Figure 1.11. For combined (vertical, inclined or horizontal) electrodes, the total length shall be considered. The minimum length may be disregarded provided that an earth resistance of the earth-termination system less than 10 Ω is achieved. The type A arrangement is suitable for low structures such as family houses, existing structures, lightning protection system with rods or stretched wires and for an isolated lightning protection system. Where there is a ring conductor, which interconnects the down-conductors, in contact with the soil, the earth-termination system is still classified as type A if the ring conductor is in contact with the soil for less than 80 per cent of its length. The use of plates as earth electrodes is usually discouraged because of the easy joint corrosion.

In the type B arrangement, the ring earth electrode or foundation earth electrode should have a mean radius r_e of the area enclosed not less than l_1 :

$$r_e \geq l_1 \tag{3.1}$$

where l_1 is represented in Figure 1.14 according to the different LPS classes and then according to the values of lightning current parameters selected for dimensioning.

When the required value of l_1 is larger than the convenient value of r_e, additional horizontal or vertical (or inclined) electrodes shall be added with individual lengths l_r (horizontal) and l_v (vertical) given by the following equations :

$$l_r = l_1 - r_e \text{ and } l_v = 0.5 \, (l_1 - r_e) \tag{3.2}$$

The number of additional electrodes connected to the ring earth electrode shall not be less than two or less than the number of down-conductors. The additional electrodes should be connected at points where the down-conductors are connected and, for as many as possible, with equal spacing.

In order to minimize the effects of corrosion, soil drying and freezing and then stabilize the value of conventional earth resistance, the ring earth electrode and embedded horizontal electrodes should be buried at a minimum depth of 0.5 m and at a distance of about 1 m around the external walls. Vertical and inclined earth electrodes shall be installed at a depth of upper end at least 0.5 m and distributed as uniformly as possible to minimize electrical coupling effects in the earth.

The type B earth-termination system is preferred for meshed air-termination systems, bare solid rock, lightning protection systems with several down-conductors and for structures with extensive electronic systems or with high risk of fire.

Interconnected reinforcing steel in concrete foundations or other suitable underground metal structures should preferably be used as earth electrodes. When the metallic reinforcement in concrete is so used, special care shall be exercised at the interconnections to prevent mechanical splitting of the concrete. In the case of pre-stressed concrete, consideration should be given to the consequences of the passage of lightning discharge currents which may produce unacceptable mechanical stresses. A foundation earth electrode comprises conductors which are installed in the foundation of the structure.

3.4 Touch Voltage and Step Voltage

The vicinity of the down-conductors of a lightning protection system outside the structure may be hazardous to life even if the lightning protection system has been designed and constructed according to the suitable rules of IEC 62305-3 (Figure 1.12). The hazard to life is reduced to a tolerable level if one of the following conditions is fulfilled:

☆ The probability of persons approaching or the duration of their presence outside the structure and close to the down-conductors is very low;

☆ The natural down-conductor system consists of several columns of the extensive metal framework of the structure or of several pillars of interconnected steel of the structure, with the electrical continuity assured;

☆ The resistivity of the surface layer of the soil within 3 m of the down-conductor is not less than 5 kΩ m; a layer of insulating material, *e.g.* asphalt of 5 cm thickness or gravel of 15 cm thickness, generally reduces the hazard to a tolerable level.

If none of these conditions is fulfilled, protection measures shall be adopted against injury to living beings due to touch voltage as follows:

☆ Insulation of the exposed down-conductor giving a 100 kV, 1.2/50 μs impulse withstand voltage, *e.g.* at least 3 mm cross-linked polyethylene;

☆ Physical restrictions and/or warning notices to minimize the probability of down-conductors being touched.

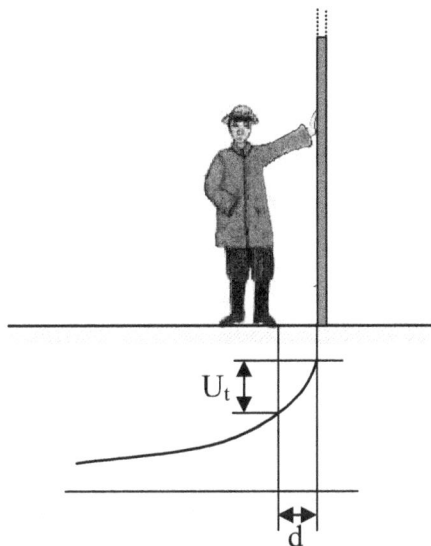

Figure 1.12: Touch Voltage.

High-current discharges can cause high earth-electrode potentials, and this can cause mortal danger to animals and human beings in the immediate vicinity. The danger increases with increasing potential difference between points of the ground which may be touched simultaneously with different parts of the body. Generally the feet are involved, hence the notion of *step voltage.* (see figure 13) When lightning strikes open ground, the current is discharged into the mass of the earth. On uniform ground, the discharge takes place in a regular pattern, and it is quite irregular when the ground is non-uniform. A person standing near the striking point is subjected to a potential difference U between the feet, such that

$$U = I \frac{\rho}{2\pi} \frac{s}{d(d+s)} \tag{3.3}$$

where, I denotes the current amplitude, ρ the resistivity, s the step length and d the distance between the striking point and the nearer leg of the person. In human beings this potential difference will cause a current to flow through the legs and lower part of the trunk.

In order to keep the step voltage as low as possible the earth electrode should be buried as deeply as feasible (at least 0.6 m deep).

Close to a down-conductor, the hazard to life is reduced to a tolerable level if one of the following conditions is fulfilled:

☆ The probability of persons approaching or the duration of their presence in the dangerous area within 3 m from the down-conductor is very low;

☆ The resistivity of the surface layer of the soil within 3 m of the down-conductor is not less than 5 kΩ m; a layer of insulating material, *e.g.* asphalt

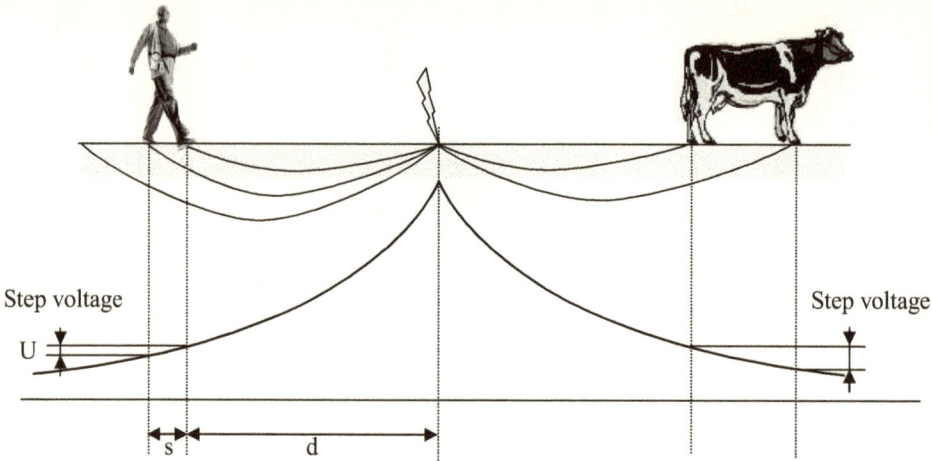

Figure 1.13: Step Voltage.

of 5 cm thickness or gravel of 15 cm thickness, generally reduces the hazard to a tolerable level.

If none of these conditions is fulfilled, protection measures shall be added:

☆ Equipotentialization by means of a meshed earth-termination system;

☆ Physical restrictions and/or warning notices to minimize the probability of access to the dangerous area within 3 m of the down-conductor.

4. Internal Lightning Protection System

Electrical and electronic systems now pervade every aspect of our lives. Permanent failure of these systems can be caused by the lightning electromagnetic impulse (LEMP) via conducted and induced surges transmitted to apparatus via connecting wiring or via the effects of radiated electromagnetic fields directly into apparatus. ([1.4]). In continuous processes, even a small transient overvoltage can cause huge financial losses.

Surges in the structure can be generated externally (surges created by lightning flashes striking incoming lines or the nearby ground and transmitted to electrical and electronic systems via these lines) or internally (surges created by lightning flashes striking the structure or the nearby ground).

Coupling of these transients into the structure happens by different mechanisms: resistive coupling (*e.g.* the earth impedance of the earth termination system or the cable shield resistance), magnetic field coupling (*e.g.* caused by wiring loops in the electrical and electronic system or by inductance of bonding conductors). Electric field coupling is generally very small when compared to the magnetic field coupling. (*e.g.* caused by rod antenna reception).

Radiated electromagnetic fields can be generated via the direct lightning current flowing in the lightning channel or via partial lightning current flowing in conductors of an external lightning protection system, according to IEC 62305-3, or in an external spatial shield according to IEC 62305-4 ([1.4]).

4.1 LPMS: LEMP Protection Measures Systems

A LEMP protection measures system (LPMS) is defined as a complete system of protection for internal systems against lightning electromagnetic pulses (LEMPs). The following basic measures are generally used in combination: earthing and bonding, magnetic shielding, line routing, and coordinated surge protective devices (SPDs). Additional protection measures include: extensions of the structural lightning protection system (LPS), equipment location and protection by isolation (use of fibre optic cables). Protection against LEMP is based on a concept of Lightning Protection Zones (LPZs) that divide the structure into a number of zones according to the threat level posed by LEMP. These zones are theoretically assigned volumes of space where the LEMP severity is compatible with the withstand level of the internal systems enclosed (see figure 14, [1.4]). Equipment is protected against lightning, both direct and indirect strikes to the structure and services, with a LPMS comprising spatial shields, bonding of incoming metallic services (water, gas, telecommunication.) and the use of a coordinated SPD network. A spatial shield is an effective screen against LEMP penetration. An external lightning protection system (LPS) or reinforced bars within the structure or rooms would constitute spatial shields.

External Zones

☆ LPZ 0_A : zone exposed to direct lightning strokes and therefore subjected to carry the full lightning current (*e.g.* roof area of a structure) and the full lightning electromagnetic field; the internal systems may be subjected to full or partial lightning surge current;

☆ LPZ 0_B : zone protected against direct lightning strokes (typically the sidewalls of a structure) but exposed to the full lightning electromagnetic field; the internal systems may be subjected to full or partial lightning surge current.

Internal Zones

☆ LPZ 1: zone protected against direct lightning strokes (typically the sidewalls of a structure); the conducted lightning currents and/or switching surges are reduced compared with the external zones LPZ 0 by current sharing and by SPDs at the boundary; spatial shielding may attenuate the lightning electromagnetic field (damped lightning electromagnetic field); this is typically the area where services enter the structure or where the main power switchboard is located;

☆ LPZ 2,., n : as LPZ1, remnants of lightning impulse currents and surge currents are further limited by current sharing and by additional SPDs at the boundary; additional spatial shielding may be used to further attenuate the lightning electromagnetic field; this is typically a screened room or the sub-distribution board area for mains power.

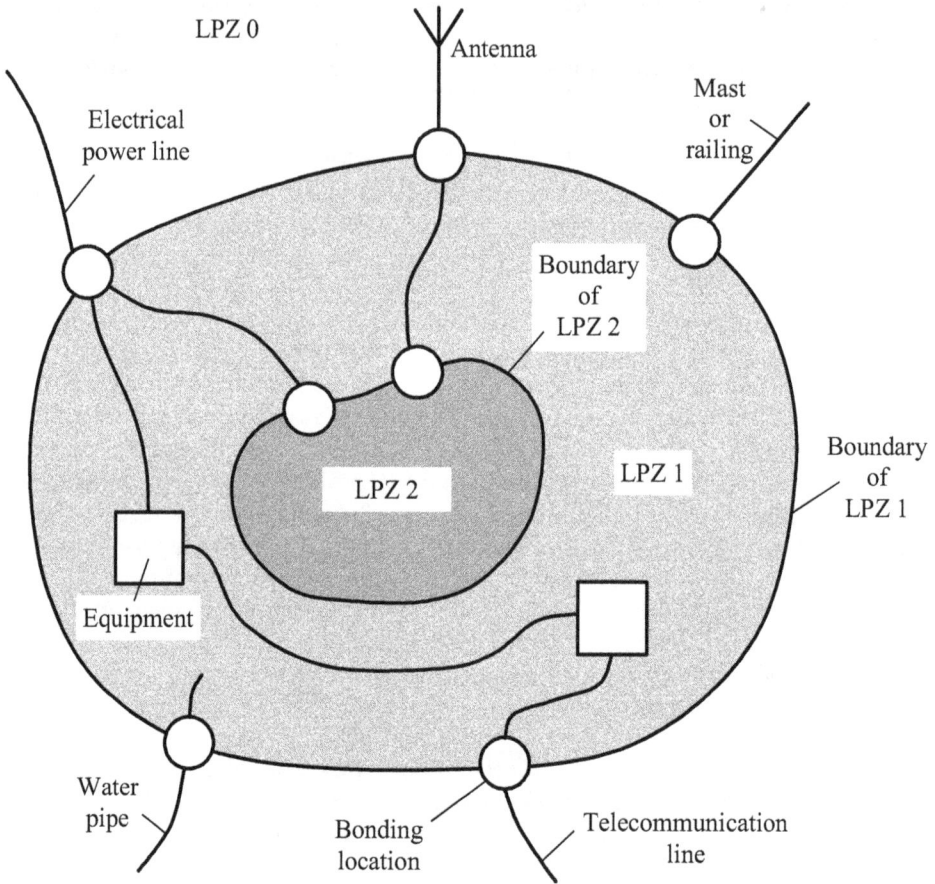

Figure 1.14: General Principle for the Division into different LPZs ([1.4]).

An example for dividing a structure into inner LPZs is shown in Figure 1.14. All metal services entering the structure are bonded via bonding bars at the boundary of LPZ 1. In addition, the metal services entering LPZ 2 (*e.g.* computer room) are bonded via bonding bars at the boundary of LPZ 2.

Earthing and bonding should always be insured, in particular, bonding of every conductive service directly or via an equipotential bonding SPD at the point of entry to the structure. Suitable earthing and bonding are based on a complete earthing system combining the earth-termination system (dispersing the lightning current into the soil) and the bonding network (minimizing potential differences and reducing the magnetic field).

In structures with electronic systems, a type B earthing arrangement (see above and reference [1.3]) is always recommended. In structures where only electrical systems are provided, a type A earthing arrangement may be used, but a type B earthing arrangement is preferable.

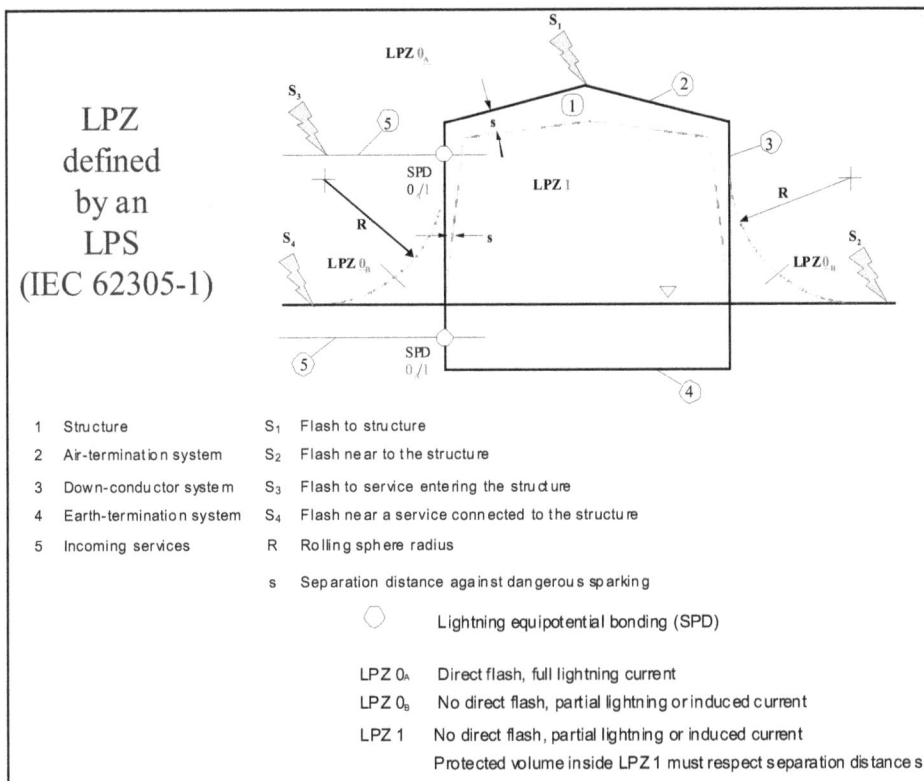

Figure 1.15: LPZ Defined by an LPS ([1.1]).

1	Structure	S_1	Flash to structure
2	Air-termination system	S_2	Flash near to the structure
3	Down-conductor system	S_3	Flash to service entering the structure
4	Earth-termination system	S_4	Flash near a service connected to the structure
5	Incoming services	R	Rolling sphere radius
		s	Separation distance against dangerous sparking

○ Lightning equipotential bonding (SPD)

LPZ 0_A Direct flash, full lightning current

LPZ 0_B No direct flash, partial lightning or induced current

LPZ 1 No direct flash, partial lightning or induced current

Protected volume inside LPZ 1 must respect separation distances

The ring earth electrode around the structure (in concrete or in open air) should be integrated with the meshed network under and around the structure, with a mesh width of typically 5 m.

The protection against LEMP to reduce the risk of failure of internal systems shall limit

☆ Overvoltages due to lightning flashes to the structure resulting from resistive and inductive coupling;

☆ Overvoltages due to lightning flashes near the structure resulting from inductive coupling;overvoltages transmitted by lines connected to the structure due to flashes to or near the lines;

☆ Magnetic field directly coupling with internal systems.

In order for the protection against LEMP to reduce the risk of failure of internal systems inside LPZ of order 1 or higher requires magnetic shields (grid-like spatial shielding or continuous metal shields or comprising the natural components of the structure itself, shielded cables, cable ducts, closed metallic cable ducts and metallic enclosure of equipment, shielding of the external lines entering the structure) to

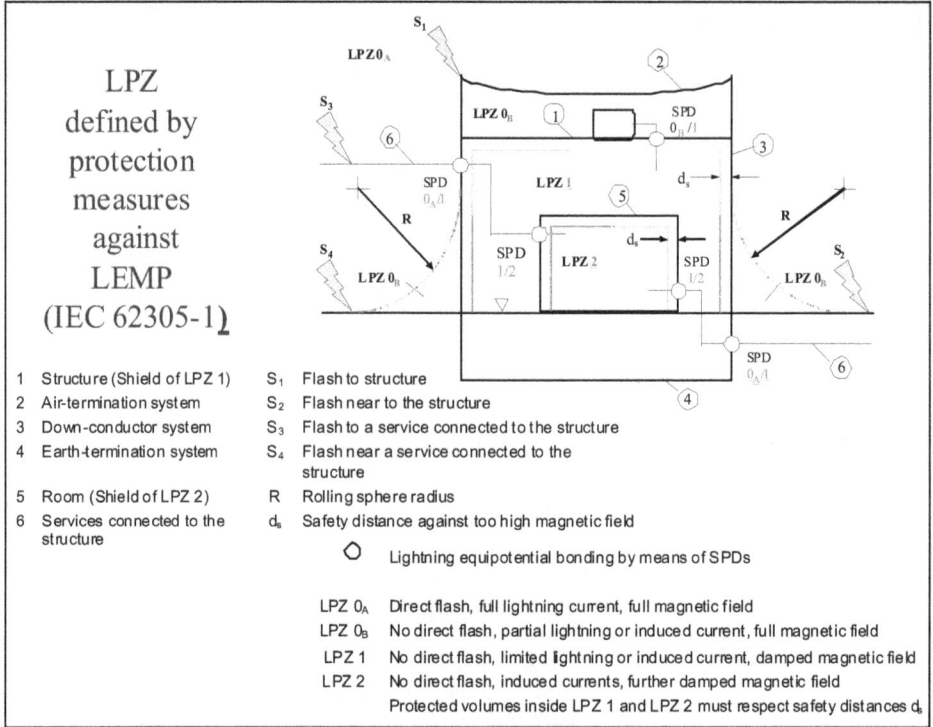

LPZ defined by protection measures against LEMP (IEC 62305-1)

1	Structure (Shield of LPZ 1)	S_1 Flash to structure
2	Air-termination system	S_2 Flash near to the structure
3	Down-conductor system	S_3 Flash to a service connected to the structure
4	Earth-termination system	S_4 Flash near a service connected to the structure
5	Room (Shield of LPZ 2)	R Rolling sphere radius
6	Services connected to the structure	d_s Safety distance against too high magnetic field

◯ Lightning equipotential bonding by means of SPDs

LPZ 0_A Direct flash, full lightning current, full magnetic field
LPZ 0_B No direct flash, partial lightning or induced current, full magnetic field
LPZ 1   No direct flash, limited lightning or induced current, damped magnetic field
LPZ 2   No direct flash, induced currents, further damped magnetic field
Protected volumes inside LPZ 1 and LPZ 2 must respect safety distances d_s

Figure 1.16: LPZs Defined by Protection Measures against LEMP ([1.1]).

attenuate the inducing magnetic field and/or suitable routing of wiring (internal lines) to reduce the induction loop and the internal surges.

At the boundary of LPZ 0_A and LPZ 1, materials and dimensions of magnetic shields shall comply with the requirements of IEC 62305-3 ([9.3]) for air termination conductors and/or down conductors.

The basic approach to the coordinated SPD protection is the same for both power lines and telecommunications lines, but because of the extensive diversity of electronic systems and their characteristics, the rules for the selection and installation of such a system are different to those which apply to the choice of SPDs for electrical systems only.

In LPMS using the LPZ concept with more than one zone, SPDs shall be located at the line entrance in each LPZ. In LPMS using LPZ 1 only, SPDs shall be located at the line entrance into LPZ 1 at least. In both cases, additional SPDs may be required if the distance between the location of the SPD and the equipment being protected is long (see annex D in IEC 62305-4, [1.4], and annex E in IEC 62305-1, [1.1]).

Internal systems are protected if their impulse withstand voltage U_w is greater than or equal to the voltage protection level U_p of the SPD plus a margin necessary to take into account the voltage drop of the connecting conductors and if they are energy coordinated with the upstream SPD.

Acknowledgements

The author would like to express his sincere thanks to his Physics staff, Pierre Lecomte and Frédéric Coquelet, who designed several figures of this presentation.

References

1. International Electrotechnical Commission, IEC 62305 standard (first edition:2006; second edition: 2010), *Protection against Lightning*, divided in four parts:

   1.1. 62305-1 : *General Principles*;

   1.2. 62305-2 : *Risk Management*;

   1.3. 62305-3 : *Physical Damage to Structures and Life Hazard*;

   1.4. 62305-4 : *Electrical and Electronic Systems within Structures*.

2. C. Bouquegneau and V. Rakov, *"How dangerous is lightning?"*, Dover Publications, N.Y., 2010; also published in French and in Chinese.

3. V. Rakov and M. Uman, *Lightning Physics and Effects*, Cambridge Univ. Press, 2003.

4. V. Cooray, *Lightning Protection*, IET, Power and Energy Series 58, UK, 2010.

5. C. Bouquegneau and B. Jacquet, *How to Improve the Lightning Protection by Reducing the Ground Impedances*, 17th ICLP, The Hague, the Netherlands, 1983.

6. R.H. Golde, *Lightning*, vol. 2, Academic Press, 1977.

7. C. Bouquegneau, P. Lecomte and L. Remmerie, *Risk Multilingual, a Complete Software to calculate the Lightning Risk for Structures*, IX International Symposium on Lightning Protection, Foz de Iguaçu, Brazil, 23rd-29th November 2007.

8. C. Bouquegneau, P. Lecomte, *Ground Impedances of Binding Loops*, Proceedings of Ground'2006 and 2nd LPE, Maceió, Brazil, November 2006, pp. 189-193.

Chapter 2

Lightning Protection of Power Transmission Lines and Distribution Transformers

Baboucarr Awe

Electrical Engineer
National Water and Electricity Company Ltd,
The Gambia
E-mail: babawe@hotmail.com

ABSTRACT

Lightning is a major cause of outages on the overhead power transmission and distribution lines and a major factor in damages to electrical equipment like transformers. Since the effect of lightning strikes on electrical overhead networks and transformers are well understood, this paper presents a historical record of outages caused by lightning on Gambia's four major transmission lines, the causes of damage to distribution transformers, the assessment done during the past three years on the four lines by the National Water and Electricity Company and the cause of transformer damage on the 11KV distribution feeders. The company had a strong base for the assessment since there has been no expansion or alteration to the four lines since they were constructed.

The assessment clearly showed how line construction can affect lightning protection. This can be ascertained on the four lines as their construction styles are different. It also showed how the use of lightning arrestors and a proper grounding system can greatly impact lightning protection on the lines and the distribution transformers on the 11KV feeders.

These conclusions were established based on earlier data on lightning related outages and transformer damages prior to the assessment and the technical interventions made on the network after the assessment. Recent data indicated that lightning related outages fell by more than 40 per cent and transformer damages fell by more than 35 per cent.

Introduction

The 33KV operated medium voltage transmission system of the Gambia is an important electrical component in off-loading the generated power from its two main power plants. Based on the geographical locations of these two power plants, the transmission lines also serve as conduits for dispatching the bulk power to various Primary Substations through the Dispatching Centre. The lines are constructed on the outskirts of the residential areas spanning from the forest areas to coastal or seaside areas. Most of these areas are prone to thunderstorms and lightning that have serious negative implications for the network stability. Due to high outage rates during the rainy season on these lines warranted thorough assessment and measures to reduce the outage rates.

33KV Lines

There are four major transmission lines within the Gambia network grid. These lines are named as follows:

Figure 2.1: 33KV Coastal Line.

a. 33KV Danida line which is 15km long
b. 33KV Brikama Feeder One (Coastal), which is 55km long
c. 33KV Brikama Feeder Two which is 72km long
d. 33KV Western Region line which is 100km long

These lines were constructed at different times and their methods of construction differed. The 33KV Danida line, shown in Figure 2.2, was constructed and commissioned in 1991 with high steel poles bolted on a concrete foundation and using horizontal line post insulators. The whole line is provided with an upper overhead shield wire with a clearance of 60cm and at an angle of 30 degrees from the first phase conductor and grounded at specific intervals. The 33KV Brikama Feeder One and Two were constructed and commissioned in 2006 with high galvanized steel poles. It mostly utilizes the Nappe Voute (NVR3) type cross arms and every pole is grounded. Glass insulators are used in all these lines. The western region line was constructed and commissioned in 2011. It also utilizes the high galvanized steel

Figure 2.2: 33KV Danida Line.

poles of about 16 to 21metres and the Nappe Voute type cross arms and every pole is also grounded. The composite type insulators are also used on this line.

Most of these lines are constructed in areas that are prone to severe weather related disturbances especially lightning and thunderstorms. A careful study to address these issues warranted a look at the historical background of all these outages.

History of Outages

Prior to the year 2010, only the three transmission lines were operational. The historical severe weather related outages particularly caused by lightning and thunderstorm during the period of August and September were observed and noted. Lightning related outages on Gambia's distribution feeders and the damages caused on transformers were also noted. Based on statistical indications and observations, the 33KV Brikama Feeder One experienced more lightning related outages than the others. The distribution feeder outages as a result of damaged cables and transformers were also noted during the above mentioned period.

Field Assessment

The outages prompted the Transmission and Distribution Department to embark on a massive field inspection of these lines to assess the root cause of the massive lightning related outages and how to address them. Based on the company's assessment and practical experience, it was realized that line construction, the use of surge arresters and proper earthing (grounding) could greatly impact lightning protection for stability purposes and preventing transformer damages. These are explained below:

On the 33KV Danida line, an overhead shield wire of about 60cm clearance, positioned at an angle of 30 degrees from the first phase conductor and grounded at specific intervals protects the line from direct lightning strikes. The insulation level of this line is 3 times the line rating. The reliability of the line is approximately 98 per cent.

On the 33KV Brikama Feeders One and Two, most of the cross arms are designed to capture any direct lightning strike and send it to ground as every pole is grounded. The grounding of these poles is done by copper wire link from the pole base to the ground rod. The insulation level on this line is 1.5 times the line rating. The average reliability of both lines is approximately 75 per cent.

On the 33KV Western Region line, most of the cross arms are designed to capture any direct lightning strikes and send the lightning current to ground as every pole is grounded. The grounding of these poles is done by a flat galvanized earth strip from the pole base to the ground rod. The insulation level on this line is 4 times the line rating and the reliability is approximately 98 per cent.

During the line assessment, the Brikama Feeder One recorded most of the severe weather related outages. This is obvious as it is constructed along the coast which is prone to lightning. It was also realized that the copper wire connecting the pole base with the ground rod on most of the poles were tampered with hence affecting the grounding on the line. It was realized that the harsh weather conditions on the

Brikama feeder one (Coastal line) such as pollution and the salinity affect the glass insulators and tend to break these insulators, thus reducing the insulation level on the lines which normally leads to more lightning vulnerability.

Regarding transformer damages on the distribution feeders, it was realized that there was inadequate or no proper grounding on most of the distribution transformers. Lightning protection devices for the transformers and underground cables were non-existent in most of our substations.

Practical Measures

From these assessments, practical measures were taken to minimize lightning related outages and transformer damages. The glass insulators are being gradually replaced with composite insulators that have higher insulation levels and are more resistant to harsh weather conditions. Surge arresters are installed every 300-600 meters and before every entry point to the primary substations. This is intended to protecting the line from switching and lightning surges. Transformer damages on the distribution feeders mostly occur because of overvoltage emanating from lightning strikes. In order to minimize the damages, the earthing (grounding) mechanism on most of the transformers on the distribution feeders were improved and proper grounding mechanisms were installed on installations that had none. The systems are tested periodically. Surge arresters were installed on most of transformer substations. These efforts were aimed at protecting the transformers and underground cables from damages caused by overcurrent emanating from lightning strikes.

Conclusion

The data shown below is typical for months of August and September in which severe weather related network outages are experienced. From observation, most of these outages occur as a result of lightning strikes and thunderstorms.

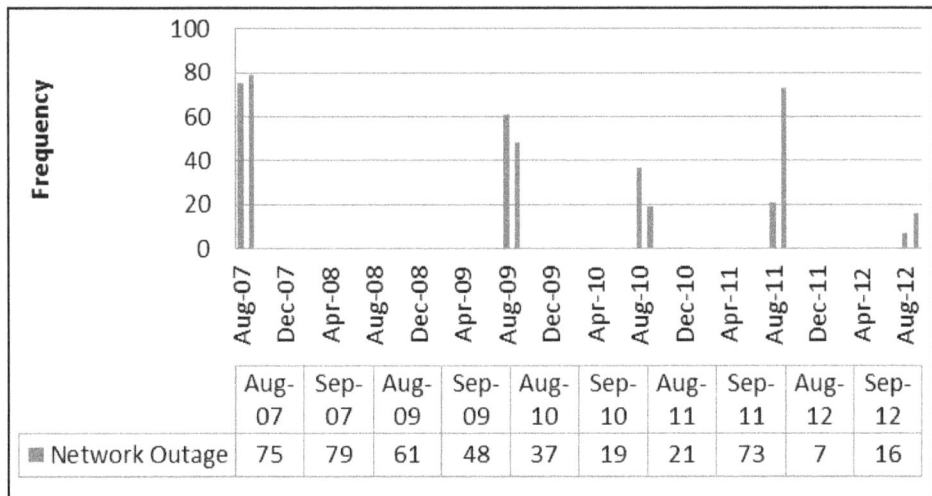

	Aug-07	Sep-07	Aug-09	Sep-09	Aug-10	Sep-10	Aug-11	Sep-11	Aug-12	Sep-12
Network Outage	75	79	61	48	37	19	21	73	7	16

Figure 2.3: Network Outage Data

Based on the data in the chart above, it can be seen that prior to 2010, weather related network outages were very high. As a result of the practical measures and interventions, a significant drop in network outages has been realized. There has also been a decrease in transformer damages. In 2012, damages to only three transformer were recorded during the rainy season compared to an average of damages to five transformer in the previous years. The assessment also revealed that the use of lightning resistant construction of increased insulation level and smaller shielding angle tend to reduce lightning transients, flashovers and high voltages impressed on the line. This can be attested to in the 33KV Danida Line.

Chapter 3

Lightning Casualties Side-Flashed in Buildings

Robert Jallan'go Akello

Professor of Electrical and Communications Engineering
MultiMedia University of Kenya, P.O. Box 30305, Nairobi, Kenya
E-mail: rjakello@mmu.ac.ke; rjakello@yahoo.com

ABSTRACT

This paper first addresses the strike paths which the lightning current from cloud to ground (CG) discharge take to either kill or injure a person. The basic reasons why human bodies are vulnerable to lightning currents are given. Thereafter, the types of buildings posing danger to persons are visited. A typical case, with pictures, where six people were killed and several injured is given. Finally, the poster on basic personal precautionary measures used in Kenya for public awareness is included. Conclusions are then drawn from experiences of the author.

Keywords: Paths, Human, Keraunomedicine, Buildings, Casualties, Precautionary measures.

1. The Lightening Current Paths[1]

Lightning strikes persons on the ground through three basic paths. First is the *direct strike*, where the channel enters the person, often the primary target in an open field, directly. This often causes casualties to herdsmen, footballers in playfields and golfers on golf courses.

Second is through the *contact voltage* either between a hand touching a directly struck object such as a tree and with feet on the ground or between spaced feet forming a step on a ground which is conducting away a lightning current. Casualties who suffer from this path are either those touching directly struck trees and such objects, while standing on wet ground or those standing astride on wet grounds such as flowing water, which is likely to conduct lightning current.

Third, and most dangerous, is a *side flash* with the lightning current entering a person near a less conducting object which has been struck. Most casualties suffer from this third path. Examples are schoolchildren standing on a veranda of a classroom whose roof has been directly struck.

2. Human Bodies

One commonality in all human bodies is that they conduct electrical currents. The best conducting system is the nervous system, with the largest concentration of cells in the brain, which controls all the body's functions. That is the reason why one is only pronounced dead when the brain is dead. The circulatory system, centred in the heart, is second since it contains blood with lots of conducting salts.

Any excess currents, such as those from lightning, will cause abnormal functions of body organs. If, for example, the heart malfunctions through either atrial or ventricular fibrillation, the brain and the other body organs such as glands, liver, lungs and many others follow. This gives the reason for checking the EEG and ECG in hospitals, for information on the electrical activities of these two vital body organs, before the others shown in Figure 3.1 are checked in keraunomedicine – the medical study of lightning casualties [2,3,4].

Other exposed vital organs which can be damaged in the human body are the skin, muscles, glands, eyes, ears and bones.

3. Buildings

Buildings house people and other objects. They can be either short or tall and have their roofs, walls, windows and other parts made of either electrically conducting or insulating materials. The conducting materials such as electrical wiring, roofs, doors, windows, re-enforcements and piping are preferred paths by lightning current.

If lightning strikes a metallic roof, it will branch to follow any of these paths to reach the ground. A person within the proximity of these metallic building parts will provide a preferred path and be side-flashed to provide a path, especially if there is a disruption of flow by a poorer conductor such as a dry block wall. Lightning heat may burn the house or expansions may destroy the building.

In order to avoid side flashing and destruction of the building, it must be provided with a professionally designed lightning protection system (LPS) [5–8] such as in Figure 3.2.

This is a conductor made of either copper or aluminium, starting with an aerial termination on the roof and a tape running along the wall to be electrically earthed on the ground. This system is professionally designed to conduct the lightning current uninterruptedly and safely to the ground, without preferring any other paths.

4. Casualities

Energies in a lightning channel have varied effects on the human body. The large electric current from lightning results in intense heat, which in turn causes

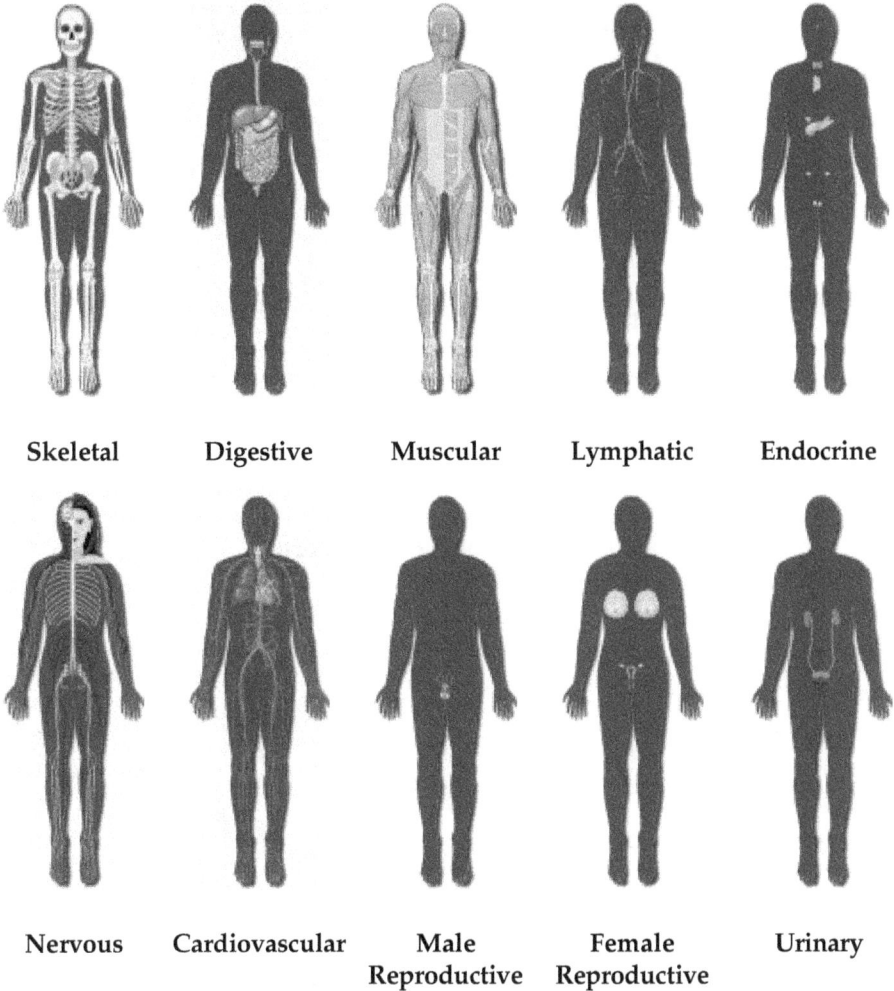

Figure 3.1: The Human Systems.

dehydration and burns along its paths. The heat also causes the air around the channel to glow, thus emitting blinding light which can cause blindness. It also makes the air to expand, resulting in explosive pressure, which displaces objects such as humans to be swept away and displaced with a strong force, causing bruises, muscle tearing and fractures. The rapid expansion of the air also causes sound which can blow eardrums and cause dumbness. Finally, the electromagnetic field from the currents cause polarization of the human cells resulting in general fatigue, blunt trauma or paralysis.

Lightning casualties who have survived death may therefore have burns, be dehydrated and fatigued, be dumb or blind, and have bruises, muscle tear and fractures from either being violently displaced by the pressure or hit by debris from a struck building. More of such damages may be seen in reference [9].

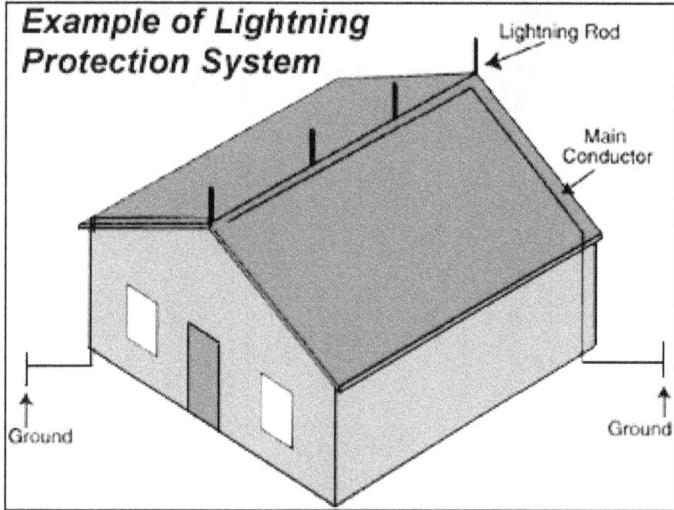

Figure 3.2a: Simple House Protection.

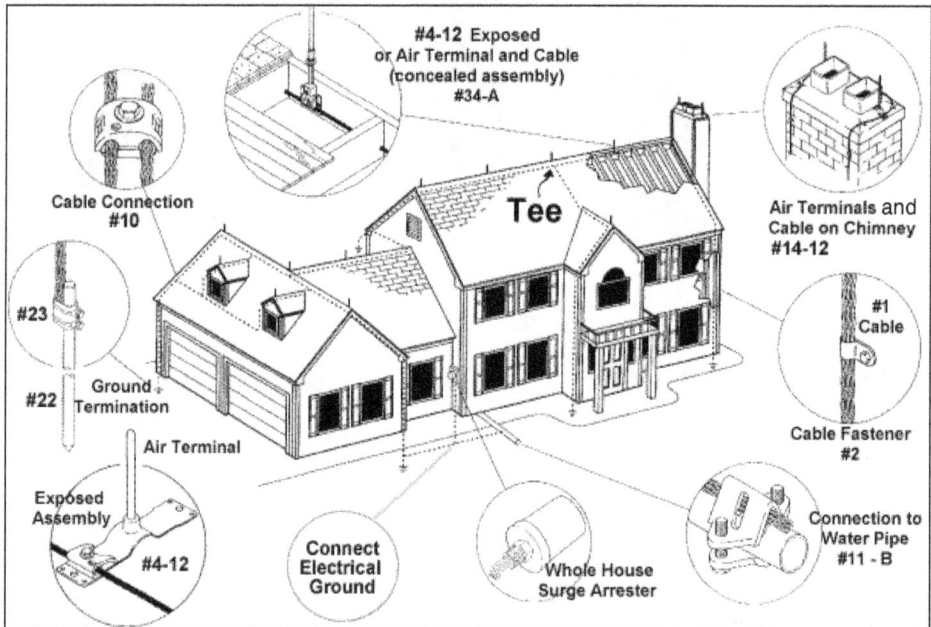

Figure 3.2b: Details of LPS Components.

Typical Case in Kenya

At about three thirty in an afternoon, lightning struck the corrugated iron roof of an incomplete church building in which many people were taking shelter from the rain. The six persons who died instantly were taken to the mortuary and the several injured to hospital. This took place in a region where people are very superstitious.

The author was invited by the Administration to investigate what might have happened. The mortuary was visited to see the bodies, accompanied by a pathologist and mortuary attendants. Next was to talk to the survivors, with an intention of

Figure 3.3a: Church Building Wall and Windows (Pictured from inside).

Figure 3.3b: Perforated Iron Sheet (See dot).

Figure 3.3c: Current Exit at Ribbon Bend (Smelted) Above Iron Ribbon.

Figure 3.3d: Lintel Shattered Along Re-Bar.

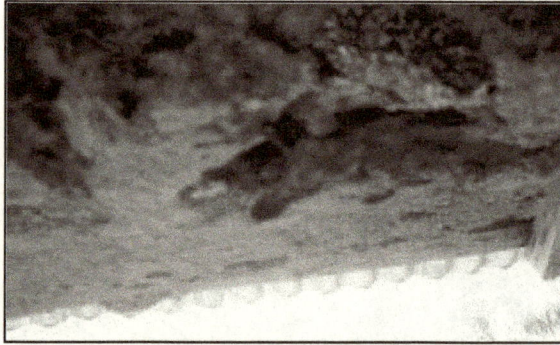

Figure 3.3e: Re-Bar Smelted by Exiting Current.

extracting the recalls, especially of their locations at the time of lightning occurrence. Finally, the scene had to be visited.

At the scene, it was observed that lightning currents arced from the roof to the metallic tie-ribbons of the roof structure, causing a small hole on the iron roof. The currents then flowed through the twisted bar re-enforcements in the lintel as in Figure 3.3.

The persons who died were side-flashed through the heads as in Figure 3.4 as they stood inside the building along the wall, below the open windows with the ribbons and re-bars above them.

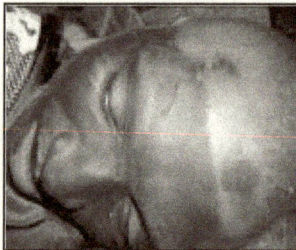

Figure 3.4a: First Fatal Victim.
(Shown multiple current entry points on head and face burns).

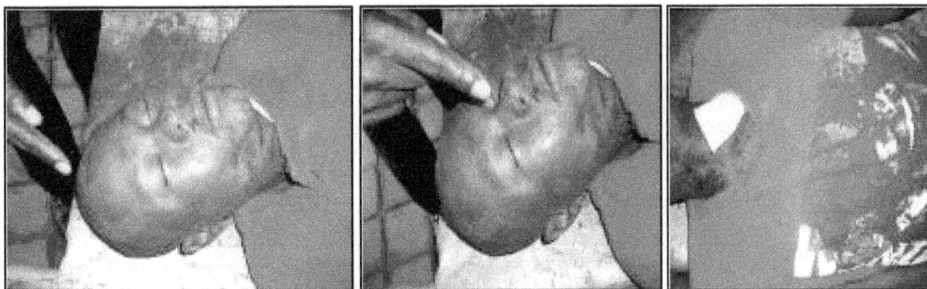

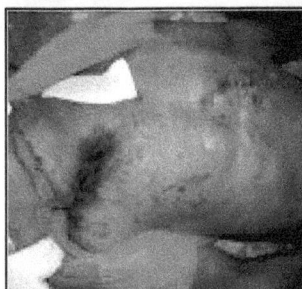

Figure 3.4b: Third Victim.
(Author pointing at current entry point. Shown burns on upper lip, cheek, neck and chest).

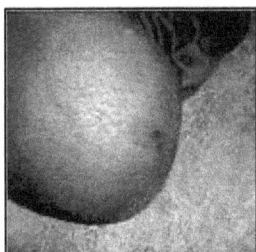

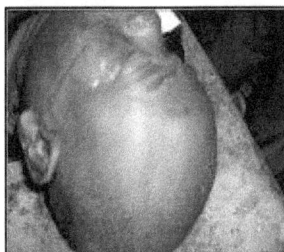

Figure 3.4c: Fourth Victim.
(Shown entry points on head, burns on face).

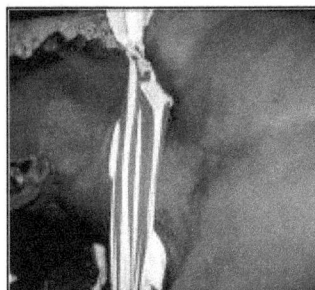

Figure 3.4d: Fifth Victim.
(Shown entry points on head, burns on cheek, neck and chest).

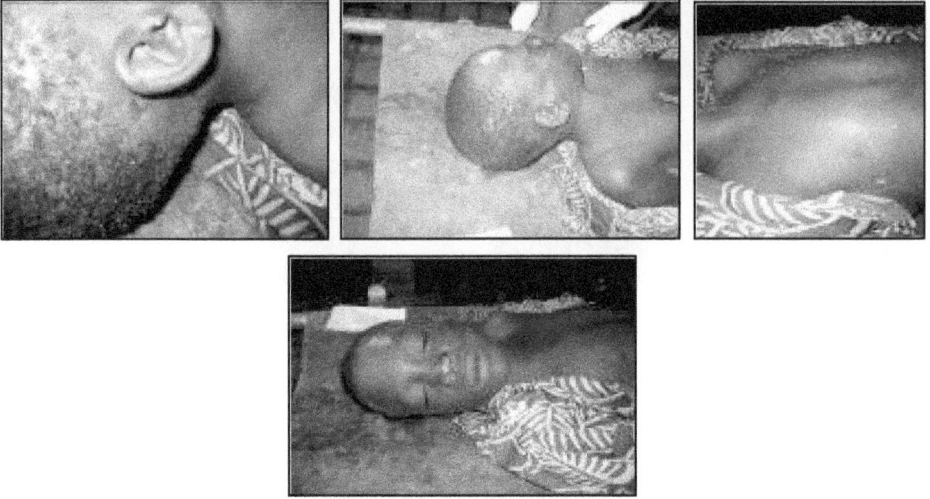

Figure 3.4e: Fifth Victim.
(Shown entry points on side of head, burns on cheek and chest).

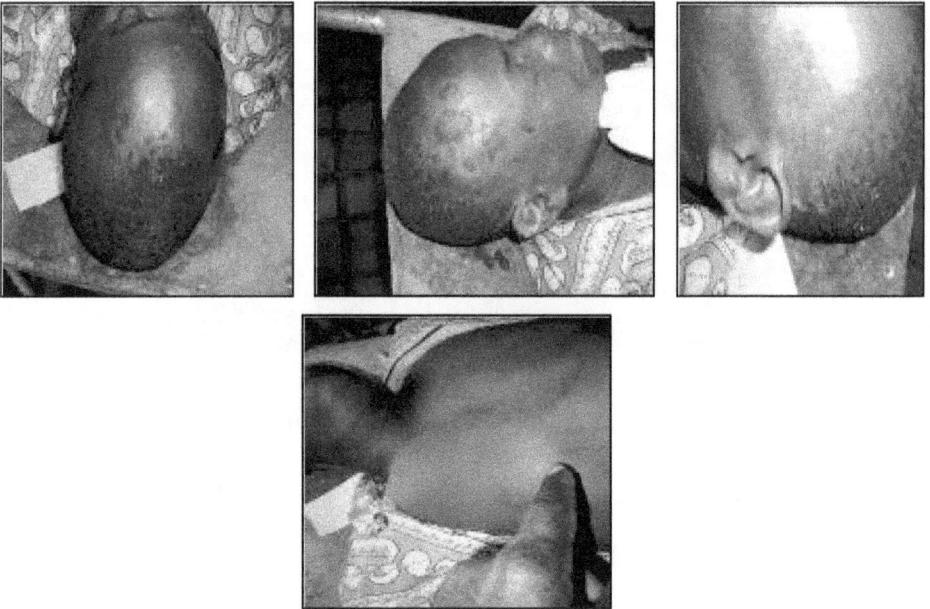

Figure 3.4f: Sixth Victim.
(Shown multiple entry points on head and ear and burns and swollen line on chest).

Those who stood adjacent to them were secondarily side flashed but survived, with traces of superficial and deep burns as in Figure 3.5; some were dehydrated and fatigued as in Figure 3.6.

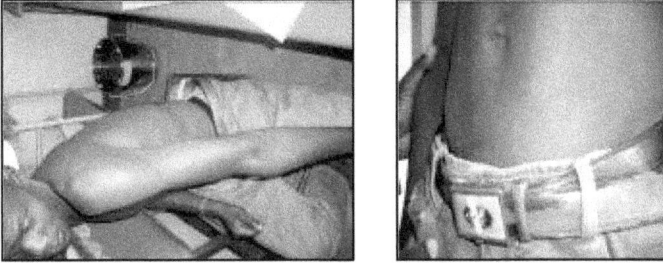

Figure 3.5a: A Flash from an Adjacent Victim Enters the Elbow Joint, Flows through the Lower Part of the Body.

(Shown current path aiming at metallic belt buckle).

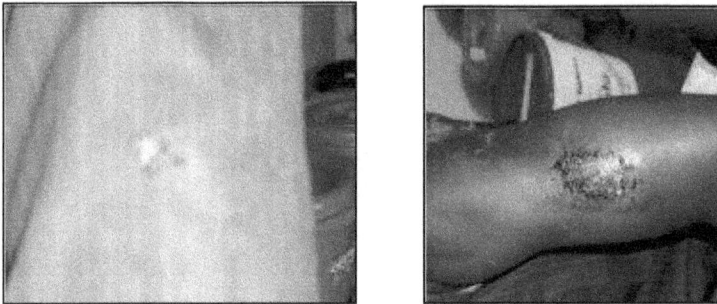

Figure 3.5b: The Flash from an Adjacent Victim.

(Shown burns the jacket penetrates to the skin).

Persons in the middle of the house who were shocked but did not get any injuries.

Figure 3.6: Traumatized, Fatigued and Dehydrated Patients.

The author finally met the bereaved residents in Figure 3.7 in order to re-assure and educate on the lightning phenomenon and personal precautionary measures.

5. Personal Precautionary Measures

In 1979, the author of this paper took an interest and got a Kenyan government licence to study lightning as a source of atmospheric radio noise. Coincidentally, the

Figure 3.7: The Author Addresses Residents.

Figure 3.8: Poster on Personal Precautionary Measures.

number of lightning casualties, similar to that illustrated above, peaked in schools and markets in the early 80s. The author was then tasked by the government to find means of alleviating the problem. This culminated into presentations [10–15] of some of the findings in different fora.

After studying the lightning patterns and statistics in different parts of the country, the poster on personal precautionary measures, show in Figure 3.8, was drawn, distributed to schools and places for public gatherings, and used at public awareness lectures in all prone districts, especially in the Lake Victoria Region.

The results, up to now, are encouraging since the incidents have reduced.

6. Conclusions

With an average annual increase in human population of 3.5 per cent and a lightning density being, according to NASA [16] up to 70 flashes per km^2 per year in Tropical Africa, there is need for an international coordinated effort in alleviating damages caused by lightning, especially the human casualties in this belt of the globe.

First, awareness on personal precautionary measures needs to be intensified since it proves to be the cheapest and easiest way to save lives. Secondly, more attention needs to be placed on Building Standards to reduce the side flash deaths and injuries. Finally, there is need to conduct post mortems on such casualties in order to give conclusive professional evidence from keraunomedical check list.

References

1. Rick Curtis, "Lightning Hazards and Safety". http://www.outdoored.com/community/w/articles/lightning-hazards-amp-safety.aspx

2. Mary Ann Cooper, "Lightning Injury Research Program". http://uic.edu.labs/lightninginjury/

3. William T. Hark, MD, "Human Effects of Lightning Strikes and Recommendations for Storm Chasers". www.harkphoto.com/light.html

4. Selected publications. http://uic.edu/labs/lightninginury/pubs.htm

5. http://www.lightningproducts.com/installation.html

6. http://www.a1 lightning.com/residential.html

7. http://stormhighway.com/protection.php

8. Google: "Images for Lightning Protection Systems"

9. http://patient.co.uk/doctor/Electrical-Injuries-and-Lightning-Strikes.htm

10. R.J.Akello, "Lightning in the Tropical Zone – The Kenyan Case Study", Proc. ICLP, Graz, Austria (April, 1988).

11. R.J.Akello, "From Research Data to Public Awareness: The Lightning Research", Proc. of National Seminar on the Teaching of Science in Kenya's Education System, Commission for Higher Education, Nairobi. (July, 1990).

12. R.J.Akello, "Lightning Protection in Kenya", Proc. IEEE AFRICON '96, University of Stellenbosch, South Africa (25-27 September, 1996).

13. R.J.Akello, "Electromagnetic Radiation Effects on Human Beings and the Environment", An Invited Paper, Proc. RADIO AFRICA '97, Nairobi, (August, 1997).

14. R.J.Akello, "Lightning Research In Eastern Africa", Proc. 18th International Conference on Lightning Detection (ILDC), Helsinki, Finland. (6th-9th June, 2004).

15. R.J.Akello and Ogada M., "Lightning Activities in East Africa" Proc. 19th International Conference on Lightning Detection (ILDC), Tucson, Arizona, USA. (24th-25th April, 2006).

16. http://geology.com/articles/lightning maps.html

Chapter 4

Implications of the Imbalance among Knowledge, Attitudes and Practice (KAP) Concerning Lightning in Malawi

Gilbert Reginald Phiri

Centre for Development Research and Information
in Southern Africa (CEDRISA)
Tukombo Girls' Secondary School, P/B 2, Kande, Nkhata Bay,
Malawi
E-mail: gr_phiri @ yahoo.co.uk

ABSTRACT

In-depth localized studies and national records show that Malawians are very conversant with the lightning phenomenon. Malawians' awareness encompasses the following key aspects: frequency and timing of lightning strikes; damages caused by lightning; rate of deaths caused by lightning strikes not to mention life-altering ailments *i.e.* memory loss, dizziness, body weakness, numbness and burns (Department of Disaster Preparedness and Management, 2009; Mulder *et al.*, 2012). However, Malawians' attitudes towards the lightning phenomenon are very diverse. They range from beliefs and myths which border on ignorance about lightning (Phiri, 2009; Mulder *et al.*, 2012) to those that are based on sound and/or working scientific knowledge regarding the physics of lightning. This paper discusses the discrepancy between knowledge (K) and attitudes (A), and how individuals and communities react (P) to lightning. To counter the discrepancy, one needs to look for strategies to empower communities should to detect signs of lightning – induced danger. Mitigation strategies need to have a heavy skew towards community education, community trouble shooting and group counselling. The paper concludes with a list of recommendations aimed at improving lightning awareness and education in Malawi.

Keywords: *Avoidance syndrome/Reaction, Community radios, Community trouble shooting, Fairy tales or riddles, Group counselling, Mitigation, Skewed, Traditional herbalists (THs).*

Introduction

Malawians are very conversant with the lightning phenomenon. Their awareness encompasses key aspects: frequency and timing of lightning strikes; damage caused by lightning; rate of deaths caused by lightning strikes over and above life–altering ailments *i.e.* memory loss, dizziness, body weakness, numbness and burns (Department of Disaster Preparedness and Management, *1946 to 2011*; Mulder *et al.*, 2012).

On the other hand, Malawians' attitudes towards the lightning phenomenon range from beliefs and myths which border on ignorance about lightning (Phiri, 2009; Mulder *et al.*, 2012) to those that are based on sound and/or working scientific knowledge of the physics of lightning. Naturally, how individual Malawians react prior to strikes by lightning *i.e.* at the onset of the rainy season (September to October)}, during the rainy season (November to March/April) and after each lightning strike is dependent on the level of awareness and willingness to accept the lightning phenomenon as a natural physical occurrence. If lightning were understood as a purely natural physical phenomenon, retrieval of the dead and treatment of survivors would suffice. Instead, reactions of the relatives of the victims are a more worrisome aftermath of lightning in most areas in Malawi. This can be ascribed to the following main factors:

☆ The variance between Malawians' knowledge (K) and attitudes (A) towards lightning on one hand and how they react (P) to the phenomenon;

☆ A lack of clarity about how to classify lightning – induced disasters (*atmospheric or environmental*);

☆ The absence of lightning studies in school and university curricula.

Analysis of KAP about Lightnining in Malawi

The discrepancy between knowledge (*K*) and attitudes (*A*), and how individuals and communities react (*P*) to the lightning is attributed to a wide range of factors. One factor is *the avoidance syndrome/reaction*. At all costs, rural communities in Malawi just like elsewhere in Africa, put interpersonal relationships at the heart of their individual and social lives. In the event that someone is struck by lightning, the barest minimum is to point an accusing finger at some suspected adversary in the community. Should a life be lost or a survivor show signs of life-altering ailments, the natural reaction is either to evict or influence the community to evict the prime suspect or group of prime suspects from the community (Phiri, 2006; Mulder *et al.*, 2012, Nyasa Times, 2012). How detrimental this attitude towards lightning is, one may ask. Although the avoidance syndrome is baseless, its social and economic implications are dire. Apart from global statements about lightning prone districts in Malawi, strikes by lightning have no boundary. Moving out or inducing the eviction of suspected adversaries is no solution at all. On the technical front, the avoidance reaction simply puts communities at greater risk. One direct result of the avoidance reaction is that communities do not strive to learn about the dangers of lightning. In fact, they frequently turn a deaf ear to any expert advice on preparedness for lightning and management in the event of lightning strikes. One expression of this scenario is that on more occasions than one, installation of solar panels and connections to the national electricity grid are made without due regard for safety measures. Although they are by far not statistically representative, a few examples of lightning strikes in recent times can be cited here. On 19th December, 2005, eleven (11) worshippers were struck to death by lightning while five (5) more were injured in the same incident which occurred in *a corrugated iron roofed prayer house* at Chiseng'eze in Mzimba West. On 18th and 19th December, 2012, two and equally tragic incidents occurred in two separate areas. In one incident, as many as twenty (20) residents of Mdyaka village, *whose houses are powered by solar batteries,* had to be rushed to Chintheche Health Centre, Nkhata Bay Central, after being struck by lightning[1]. In the second incident, two people were struck to death on a road. In another incident, two tea pickers were struck dead in the field {*Kawalazi Tea Estate, Nkhata Bay Central*, in mid December, 2012 *(personal observations)*}. In yet other cases that occurred earlier or later, similar fatalities were registered. As will be noted later, loss of life and the human injuries caused by lightning were blamed on the work of devilish/jealous neighbours. No one thinks of putting the blame of the lightning strikes on developers who disregard the installation of protective earth (copper) wires to the prayer house or the affected houses.

On the national front, Government continues to portray lightning simply as one of many natural phenomena which cause death or injury to people and damage to property (Department of Disaster Preparedness and Management, *1946 to 2011*). Mulder *et al.* (2012) share this view by downplaying the impact of lightning strikes. According to the authors, lightning strikes can easily *"deflect attention away from more prominent causes of mortality in rural areas."* Downplaying the impact of lightning has equally dire consequences. The heart of the matter is that there is no attempt at providing basic and/or civic education to communities in Malawi about the danger lightning poses to human life. This is emanating from the fact that the precise responsibility for lightning education is not clear. Is lightning an environmental issue or simply a severe weather condition? On the other hand, is one justified to treat lightning education as a subject for academic discourse or something that can casually be thrown to providers of civic education [*i.e.* the National Initiative for Civic Education (NICE) or the Ministry of Information and Civic Education]? Alternatively, should lightning education be left in the ambit of health surveillance personnel? Incidentally, if one cast his or her net farther out, would he or she be justified to leave lightning education to the media fraternity or non-governmental organizations such as those

---

1 The solar panels in Mdyaka village were supplied and installed by a foreign donor institution whose personnel should have been adequately informed about electricity installation precautions.

that specialise in rural living or indeed to the police (Tayanja – Phiri, 2005; Khunga, 2012; MANA, 2012; Nyasa Times, 2012; Malikwa, 2012)? [2] The list of institutions that should have competence in lightning education is much longer than what has been presented here. However, no single institution has currently assumed the responsibility of educating Malawians about the nature and dangers of lightning; let alone how to manage lightning strikes.

Where does this leave communities, especially those in rural areas? From the foregoing paragraph, the need for Government to streamline lightning education is clear. The omission of this service leaves communities at a loss. In fact, the omission has resulted in a vacuum of vital information about the nature and danger of lightning. This information vacuum highlighted here, has entrenched KAP in the lightning phenomenon in Malawi. One likely result of the entrenchment of KAP in lightning is communities' decision to fend for themselves by looking up to anyone they deem helpful. The majority of them consult traditional herbalists (THs) in their shrines to seek solace (Phiri, 2012). This worsens the situation even more especially when and where lightning victims develop life – altering ailments *i.e.* memory loss, dizziness, body weakness, numbness and burns. Since lengthy periods are spent at the shrines of traditional herbalists (THs) medication in hospitals is sought as the last resort.

While accepting that there are causes of mortality in rural areas that claim more lives than lightning does annually, everyone who is concerned about human health and safety should not lose sight of the trauma that lightning strikes cause. Every strike by lightning is sudden. When large numbers of people are struck at once (*as noted in this paper*), entire communities are traumatized. In the event of death or the development of life – altering ailments, victims and their relatives are irreversibly adversely affected.

Mitigation Measures

The National Geographic Society (1996 – 2012) has classified lightning as a common phenomenon of extraordinary power such that a single bolt can produce up to one billion volts of electricity. The one billion volts produce heat that is five times that felt on the surface of the sun. What is even more is that, lightning bolts are unpredictable; hence human reaction to lightning strikes is circumstantial.

A snap survey reveals that there is no one prescription on how communities should manage lightning – induced disasters. One would have wished to have communities empowered to detect signs of lightning as well as developing mitigation strategies for it. If lightning is treated as one of the natural disasters for which mitigation strategies have to be developed, community education, community troubleshooting and group counselling top the list of measures that could reduce the large KAP gap discussed earlier.

Community education in Malawi takes many forms. The most common one is community troubleshooting. This strategy is one in which whole villages set out to investigate causes of problems of any kind. In their quest to identify solutions to

---

2   All stories on lightning covered by local print media are based on police reports.

problems, most communities consult traditional herbalists (THs). Whatever advice they source from THs is circulated among all members of the community. However, the advice thus obtained is of limited use because it has a heavy skew towards myth, belief and fear.

Once they have arrived at what they consider to be the most plausible cause of their problems, affected community members are comforted through group counselling. Group counselling usually starts with or includes vigils. Selected members or whole communities spend considerable lengths of time around the homes of victims or their relatives to comfort them. Malawi is blessed with community radio stations. The radios cover a wide range of community – specific subject areas. To date, however, no single community radio has taken up lightning as a phenomenon that requires constant discussion.

An additional mode of information dissemination in Malawi is the open classroom (Phiri, 2009). Known as *sangweni or mphala* (for men) and *kuka or nthanganeni* (for women), the open classroom is a forum where elders hand down knowledge, attitudes, beliefs, myths and skills through fairy tales (*virapi*), riddles (*nthalika or mikuluwiko*), songs, jingles and direct apprenticeship. With the exception of apprenticeship, all the modes listed here call for elders to gather youths to tell the latter about their experiences partly to while away time and partly to prepare the youths for adult life.

Conclusions

From the foregoing, it is clear that KAP in lightning has dire consequences. Large numbers of people in Malawi are taken unaware by lightning strikes because of many factors such as, superficial knowledge concerning lightning, the great variance between natural causes of lightning and peoples' beliefs and the lack of clarity on who is responsible for lightning education.

The following recommendations are therefore made:

☆ The Government of Malawi should establish a Centre which should coordinate research programmes and information dissemination on lightning.

☆ The Government of Malawi should establish an inspectorate which should be charged with the responsibility of enforcing development standards with special emphasis on safety measures.

☆ Personnel working for community radios should be trained to enable them to start handling lightning education.

☆ Journalists and police should be trained to enable them to jointly and/or separately compile comprehensive reports on lightning.

☆ Providers of civic education should start embracing lightning as a field needing a lot of attention.

☆ Health surveillance officers should prioritize lightning as much as they do with the traditional prominent causes of death.

References

1. Arora, Ravindra and Gomes (2009)(Editors): *The Lightning Phenomenon: Need for Awareness, Detection and Protection from Danger caused by Lightning* Delhi Daya Publishing House.

2. Department of Disaster Preparedness and Management: *Natural disasters in Malawi – 1946 to 2011*. Lilongwe, Malawi.

3. Khunga, Suzgo (2012, December 19): *Lightning kills 3 in Balaka* The Daily Times National p. 1.

4. Malikwa, Mercy (2012, December 31): *Lightning kills 2, injures 4 others* The Nation, National p. 3.

5. Malawi News Agency (MANA) (2012, December 19): *Lightning kills 2 in Ntcheu* The Nation National p. 1.

6. Mulder, Monique Borgerhoff; Lameck Msalu; Tim Caro, and Jonathan Salerno (2012): *Remarkable Rates of Lightning Strikes Mortality in Malawi*. PLoS ONE 7 (1): e29281. Doi: 10.1371/journal.pone.002981

7. Nyasa Times (2012): *Another lightning kills two, seriously injures one in Karonga* http://www.nyasatimes.com

8. Phiri, Gilbert R. (2006): *Human rights and gender – based violence in Malawi* Mzuzu University, Malawi (unpublished).

9. Phiri, Gilbert R. (2009): *Prospects for and Constraints to Effective Environmental Education in Malawi in* Neil Taylor, Michael Littledyke, Frances Quinn and Richard K. Coll (Eds): Environmental Education in Context – An international perspective in Health Education in Schools and Local Communities Rotterdam/Boston/Taipei, Sense Publications.

10. Phiri, Gilbert R. (2012): *Lifting the lid on HIV, AIDS and Tuberculosis in Malawi* in Neil Taylor, Michael Littledyke, Frances Quinn and Richard K. Coll (Eds) Health Education in Context – An International Perspective in Health Education in Schools and Local Communities Rotterdam/Boston/Taipei, Sense Publications.

11. Tayanja – Phiri, Francis (2005): *Lightning kills 11 people in Mzimba* The Nation, National p. 1.

12. The National Geographic Society (1996 – 2012): *Lightning.*

Chapter 5

Protection against Lightning : Standards and Applications

Hasbi Ismailoglu

Kocaeli University Engineering Faculty
Electrical Engineering Department, High Voltage Laboratory
Umuttepe Yerleskesi 41380 Izmit, Kocaeli, Turkey
Telephone : (+)90.262.303 35 01, Fax : (+)90.262.303 30 03
E-mail: hasbi@kocaeli.edu.tr, hasbi41@gmail.com

ABSTRACT

Lightning strikes can lead to deaths, and cause significant losses and damages to property. Therefore, some measures must be taken to protect the structures against lightning strikes. Although international standards related to lightning protection have been published, important differences are encountered in practice. These differences are mostly due to proposals of some national standards to use the non-conventional protection systems. In this study, the effects of lightning strikes on structures were summarized and the standards related to lightning protection were examined. On the other hand, the problems caused by widely used non-conventional systems are discussed.

Keywords: Lightning strike, Protection against lightning, Touch voltage, Step voltage, Potential funnel, Conventional protection system, Non-conventional protection system.

Introduction

Lightning is one of the exciting and interesting yet dangerous events of nature. The number of lightning strikes that occur in different geographical locations of the world is over 1800 per minute, at different frequencies and amplitudes [1]. For example, lightning does not occur, on the north and south poles of the world; whereas in the equatorial region of the world, lightning strikes occur on most days of the year. Lightning strikes to structures or to areas near the structures are hazardous to people, to the structures and to their contents and installations. On the other hand, in rural areas direct lightning strikes cause a significant number of life losses.

To prevent or at least reduce these losses due to lightning strikes, protection measures against the lightning strikes must be taken. In rural areas it is important to inform the public on simple measures that could be implemented to prevent or to minimize the death risks.

There are no devices or methods capable modifying the natural weather phenomena to prevent lightning discharges. Lightning discharges can be explained by the streamer theory. If the electric field intensity grows enough due to accumulation of charges in the cloud, discharges may occur in the cloud, between the clouds or between the cloud and the earth (lightning). The characteristic properties of a lightning current may be defined by the polarity, time parameters, waveform and the peak value of its magnitude.

The aim of lightning protection is to eliminate or minimize the direct and/or indirect effects of lightning strikes. The importance of lightning phenomena in the past was mainly concentrated on its effects on human beings and on the resulting fires. However, technological developments and daily living standards have brought new concepts and resulted in some improvements on protection schemes. On the other hand, to ensure absolute protection against the direct or indirect effects of lightning is generally known to be very difficult and too expensive. For example, when the lightning strikes the lightning protection system or a structure, high potential differences may occur in the transition regions of the ground and/or around the structure, due to the grounding system and magnitude of the current. The resulting potential difference may lead to formation of dangerous step and touch voltages, in the immediate vicinity of the object in question.

Effects of Lightning Strikes

Lightning strikes result in strong effects such as heat, thermodynamics, electrodynamics, electrochemical, light, sound and electromagnetic field. These effects may cause severe damages to physical structures and losses of human and animal lives.

The effects of lightning strikes can be examined in different groups, such as direct effects, near field and far field indirect effects. Direct effects are caused by current transfer via direct attachment of a lightning strike. Near field indirect effects depend on the transients due to lightning currents and to electrical fields. In this way, induced voltages may cause a breakdown in system insulation and of the air, and may produce sparking which could be hazardous. Far field indirect effects occur when the structure acts as a receiving antenna, being in the far field of the lightning channel which is acting as a transmitter. These effects of induced transients are similar to those mentioned above but less hazardous because of the much lower intensity [2].

Direct attachments of lightning strikes may cause thermal damage and disruptive mechanical forces due to ohmic heating, and arc root damage at attachment points. They also produce acoustic shock waves, magnetic pressure and forces, sparks and exploding arcs. The sparks may be electrical sparks or thermal sparks, and may occur either separately or together. As an example, damage caused accidentally on an epoxy coated hardened concrete floor of a high voltage laboratory, by an impulse current of 8/20 μs and approximately 5 kA in amplitude, is shown in Figure 5.1.

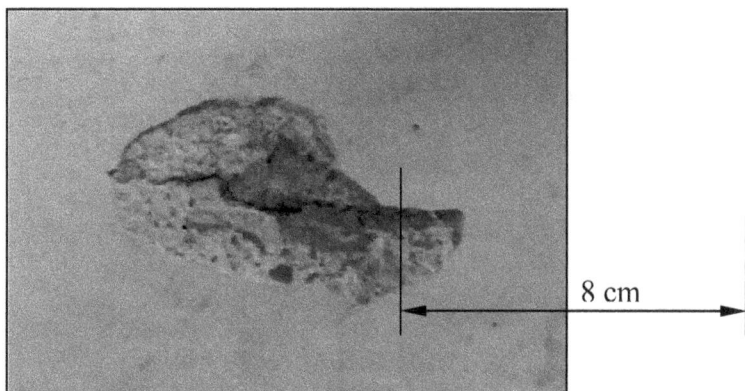

Figure 5.1: The Damage Caused by an Impulse Current of 8/20 μs and Approximately 5 kA in Amplitude on the Floor of the Laboratory.

In addition to the magnetic pressure on an isolated conductor, there are interactive forces between two adjacent conductors carrying current. The force is proportional to the product of the currents and inversely proportional to the distance between them [2, 3]. This force can be approximately calculated using the following equation:

$$F(t) = \frac{\mu_0}{2\pi} \times i^2(t) \times \frac{l}{d} \tag{1}$$

where,

$F(t)$ is the electrodynamic force (N);

$i(t)$ is the current (A);

μ_0 is the magnetic permeability of free space (vacuum) ($4\pi \times 10^{-7}$ H/m);

l is the length of conductors (m);

d is the distance between the straight parallel sections of the conductor (m).

The force may be strong enough to deform and break off the conductors. Similar force acts also on the conductor forming an angle of 90°.

On the other hand, when the lightning strikes a protection system or a structure, high potential differences may occur on the structure and in the transition regions of the ground. The variation of the potential differences (potential gradients) can be represented by a potential funnel, as shown in Figure 5.2. The amplitude of the potential differences depends on the current amplitude and on the equivalent impedance of the system. The potential differences can be approximately calculated using the following equation:

$$V(t) \cong R \times i(t) + L \times \frac{di}{dt} \qquad (2)$$

where,

   R is the equivalent resistance of lightning current path;

   $i(t)$ is the lightning current;

   L is the equivalent inductance of lightning current path;

$\frac{di}{dt}$ is the slope of the rising front of the lightning current pulse.

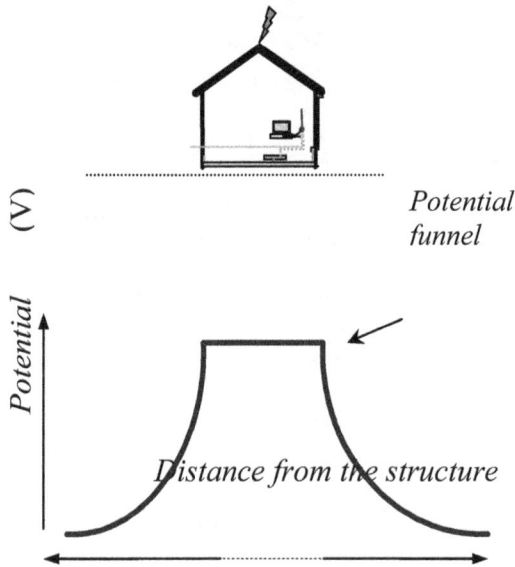

Figure 5.2: The Variation of the Potential Differences with Respect to Distance from the Structure, at Instant of Lightning Strike.

The potential gradient can cause sparks between accessories inside the structure and can lead to dangerous step and touch voltages outside the structure as well as within the structure. In addition, over-voltages may occur and spread to the environment of the structure through the communication and power lines, in the form of traveling waves.

Each type of damage to a structure alone or in combination with others, may produce different consequential loss. The type of loss that may appear depends on the characteristics of the structure itself. Some of them can be summarized as follows.

Effects of Lightning on an Unprotected Structure Made of Non-Conductive Materials

The types of damages and losses are fully dependent on the material used to construct the unprotected structure and on the severity of lightning currents. Some of expected damages and risks are listed below, according to the type of materials used in construction:

☆ Combustion of flammable materials, such as wood, mud brick etc., especially at the attachment point of the strike,

☆ Explosive fractures on non-conductive materials, such as bricks and briquettes,

☆ Formation of potential gradients (potential funnel) and of dangerous step and touch voltages,

☆ Electrical discharges in devices and between different accessories, within the building,

☆ Formation of over-voltages, spreading through lines in the form of traveling waves.

Effects of Lightning on an Unprotected Structure Made of Reinforced Concrete

Some of expected damages and risks on this kind of structures may be listed as follows:

☆ Explosive fractures at the attachment point of the strike, and on connection points of armatures with poor conductivities,

☆ Formation of potential gradient (potential funnel) and of dangerous touch and step voltages (much less than that above),

☆ Electrical discharges in devices and between different accessories, within the building,

☆ Formation of over-voltages, spreading through lines in the form of traveling waves.

Lightning Protection Systems

The lightning protection systems may be considered in two separate groups namely external and internal systems. The first group consists of protection measures to reduce physical damage and life hazard in related structure. The second group is generally related to protection measures to reduce failures of electrical and electronic systems in the structure.

The purpose of an external lightning protection system is to protect people and property from hazards caused by lightning strikes. The system consists of the following three basic parts which provide the conductive paths, required for lightning currents:

☆ Air termination devices on the roof and other elevated locations,

☆ A grounding system,

☆ A conductor system connecting the air termination devices to the grounding system.

However, the system also requires equipotential bonding, in order to ensure safety from possible secondary effects of lightning strikes, such as side-flashes and sparks. Surge protective devices must also be provided to protect power lines and associated equipment from both direct discharges and induced currents.

The external lightning protection systems can be divided into two categories, namely, conventional and non-conventional lightning protection systems.

Conventional Lightning Protection Systems

Conventional systems use Franklin rods as air termination devices. Many decades of experience show that by combining Franklin rods located at critical points on a structure with a proper down conductor and grounding system, the damage due to lightning strike could be reduced significantly [4, 5].

The implementation of the system is described in detail in relevant standards, recommending the use of conventional systems [3, 6]. One of them is a series of standards published by the International Electrotechnical Commission (IEC). The first part of this series provides general principles to be followed for protection of structures against lightning, including their installations and contents, as well as persons [3]. It also explains the need and economic justification for lightning protection. The second part of the series addresses risk assessment for a structure due to lightning flashes to earth [7]. The third part of the series deals with the protection, in and around a structure, against physical damage and injury to living beings due to step and touch voltages [8]. The last part of the series is related to internal lightning protection systems and provides information on protection measures to reduce the risk of permanent failures of electrical and electronic systems within structures [9].

In summary, this standard series defines a conventional protection system which consists of Franklin rods and the mesh method, and emphasizes the need to create equipotential bonding and to use protection devices against current and voltage surges.

Some examples of air terminations and bonding conductors are shown in Figures 5.3 and 5.4. It should be noted that both the bonding and down conductors used in the systems are in the strip form. Conductors of this type exhibit smaller impedance to the lightning currents than those with circular cross sections, due to their lower inductances and skin effects. This contributes to a reduction in the amplitude of the potential funnel. However, the installations of these systems have some objectionable spots.

Serious drawbacks that exist in given examples are as follows:

· A bend of a conductor must not form an included angle of less than 90^0,

· A bend of a conductor must have a radius greater than 20 cm,

· At least two separate down conductors must be provided on any kind of structure, including stacks and steeples.

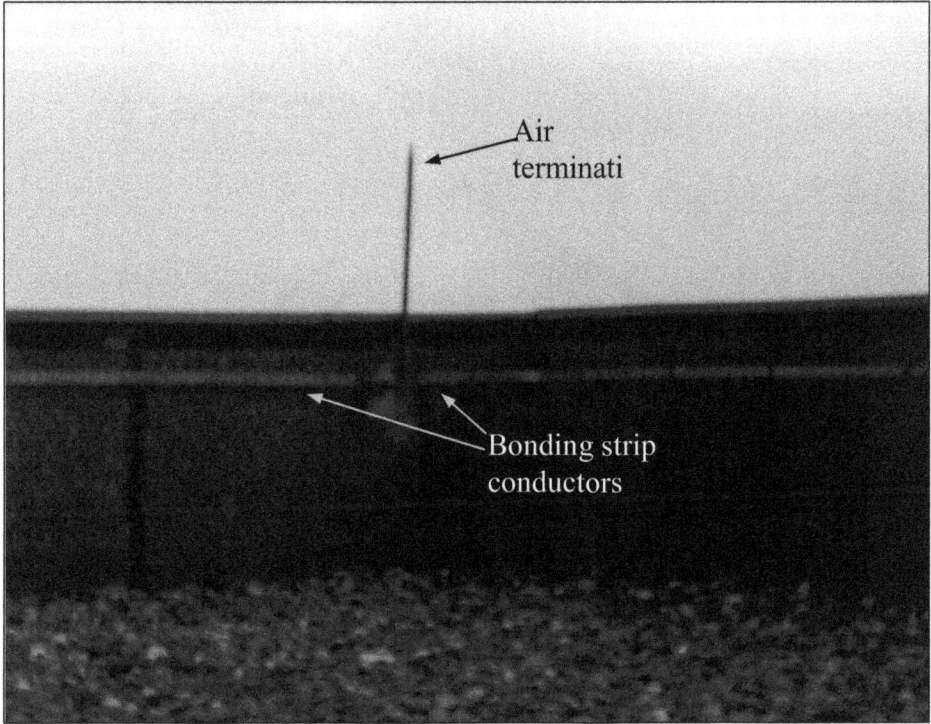

Figure 5.3: A View of an Air Termination and Bonding Conductors on a Flat Roof.

Non-Conventional Lightning Protection Systems

Non-conventional lightning protection systems include dissipation arrays or early streamer emission (ESE) rods. The idea of dissipation arrays is to utilize the space charges generated by one or several grounded arrays of sharp points to neutralize the charge in thunderclouds and thus prevent lightning from striking a protected structure. However, theoretical and experimental studies show that the dissipation arrays cannot dissipate an imminent lightning strike [10, 11].

Furthermore, experimental and theoretical investigations find results also that are in conflict with the claimed performance of ESE devices. According to these investigations, their performances are sometimes worse than those of conventional ones [4, 12]. Nevertheless, in some countries relevant standards or regulations that allow the use of non-conventional lightning protection systems are in force and therefore the non-conventional systems are widely used in these countries.

The establishment of non-conventional systems is very simple. For this reason, the installation of non-conventional systems has become a tradition in countries where their usage is allowed. For example, these systems are often used in gas stations (Figure 5.5). The remarkable point is that nearly all of the systems use paired conductors with circular cross sections as down conductors as shown in Figure 5.6. In this case,

Figure 5.4: A View of an Air Termination and Bonding Conductors on a Pyramidal Section of a Roof.

the violent lightning currents may pluck the conductors making them more vulnerable to serious hazards. Other drawbacks observed in these applications are as follows:

☆ The grounding systems are made very simple due to cost constraints so, the grounding resistances may be very large,

☆ In most cases, the systems are not equipped with equipotential bonding systems.

On the other hand, one of the documents (NF C 17-102) leading to the use of non-conventional systems was published in France [13]. However, the standard series of IEC 62305 which defines the use of conventional systems was also published in France [14-17].

Figure 5.5: An Example of Commonly Used ESE Air Termination Applications.

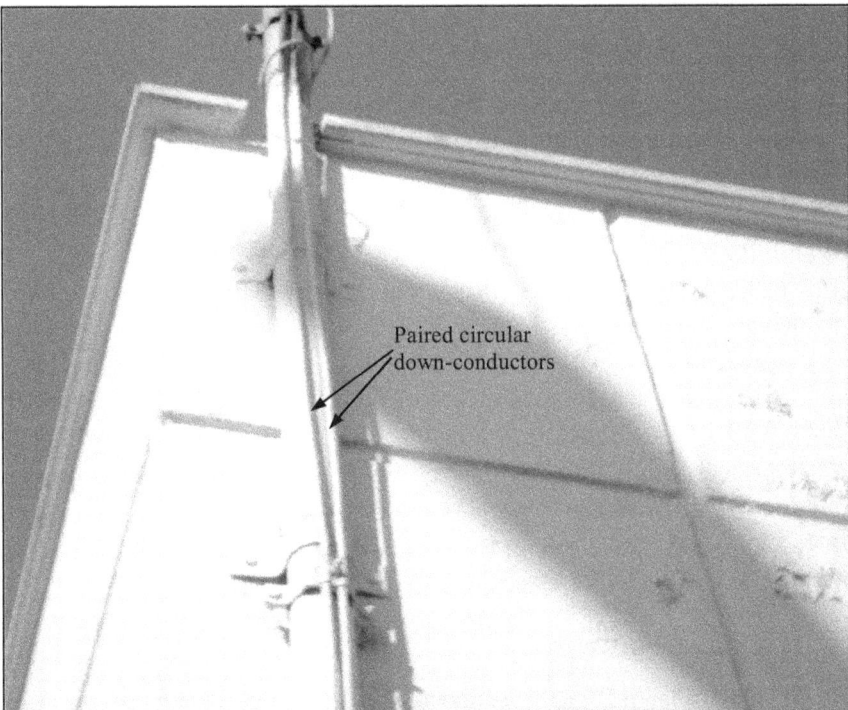

Figure 5.6: A View of Paired Down-Conductor System which Generally Used in Non-Conventional Systems.

However, even if one assumes that non-conventional systems are really advantageous, their installation schemes contradict the purpose of protection. Assuming that the claim made regarding non-conventional rods are true, since the formation of the potential difference is inevitable during lightning strike, the rod must be placed away from and not on the structure to be protected.

Conclusions

The consensus of the scientific literature, field testing, etc., is that conventional lightning protection systems are highly effective when properly installed and adequately maintained. The designation of lightning protection schemes is defined in relevant standards.

The need and economic justification for lightning protection systems must be evaluated according to the risk assessment for a structure. The establishment of a reliable protection system requires installation of equipotential bonding and of surge protection devices against current and voltage surges. Also, at least two down conductors shall be provided on any kind of structure, including stacks and steeples.

Theory and experiment show that non-conventional systems don't have any advantage over conventional systems. On the other hand, the installation schemes of these systems contradict the purpose of protection. Assuming that the claim on non-conventional rod is true, since the formation of the potential difference is inevitable during lightning strikes, the rod must be placed away from and not on the structure to be protected.

References

1. Measured Lightning Activity Worldwide, NLSI-National Lightning Safety Institute, "Structural Lightning Safety". (http://www.lightningsafety.com/index.html).

2. O. Gam, Effects of Lightning on Assets, Facilities and Structures, NLSI-National Lightning Safety Institute, "Structural Lightning Safety".

3. IEC 62305-1/February 2011, Protection against lightning – Part 1: General principles. (International Electrotechnical Commission).

4. V. Cooray, "Non Conventional Lightning Protection Systems", (Invited Lecture), 30th ICLP- 30th International Conference on Lightning Protection, Cagliari, Italy, Sept. 2010.

5. C. Bouquegneau, A Critical View on the Lightning Protection International Standard, ICLP 2004 meeting on the subject.

6. NFPA 780/2000, Standard for the installation of lightning protection systems, (National Fire Protection Association), (re-issued in 2011).

7. IEC 62305-2/February 2011, Protection against lightning – Part 2: Risk management.

8. IEC 62305-3/March 2011, Protection against lightning – Part 3: Physical damage to structures and life hazard.

9. IEC 62305-4/February 2011, Protection against lightning – Part 4: Electrical and electronic systems within structures.

10. V. Cooray (On Behalf of CIGRE Working Group C4.405), Lightning Interception - Non Conventional Lightning Protection Systems, Electra, No. 258, pp.36-41, October 2011.

11. Z. Faydali, Protection Against Lightning: Lightning Protection Systems, Their Advantages and Disadvantages, Ms. Thesis, Istanbul Technical University, Graduate School of Science, Engineering and Technology, Turkey, 2009 (in Turkish).

12. V. A. Rakov, "Lightning Discharge and Fundamentals of Lightning Protection", Journal of Lightning Research, Vol. 4, pp. 3-11, 2012.

13. NF C 17-102/July 1995, Lightning protection – Protection of structures and open areas against lightning using early streamer emission air terminals (re-issued in 2011).

14. NF EN 62305-1 (01/06/2006), Protection contre la foudre - Partie 1: Principes généraux.

15. NF EN 62305-2 (14/12/2012), Protection contre la foudre - Partie 2 : Evaluation des risques.

16. NF EN 62305-3 (14/12/2012), Protection contre la foudre - Partie 3 : Dommages physiques sur les structures et risques humains.

17. NF EN 62305-4 (14/12/2012), Protection contre la foudre - Partie 4 : Réseaux de puissance et de communication dans les structures.

Chapter 6

Lightning Hazard Mitigation in Uganda

Ahurra Kulyaka Mary[1], Chandima Gomes[2]
and Richard Tushemereirwe[3]

[1]National Meteorological Centre, Entebbe, Uganda
[2]Centre of Excellence on Lightning Protection, UPM, Malaysia
[3]State House, Uganda
E-mail: [1]ahurramary@yahoo.com; [2]chandima.gomes@gmail.com;
[3]richadt2002@yahoo.com

ABSTRACT

This paper addresses a burning issue in Uganda; the mitigation of rapidly increasing lightning hazards in the country. By analyzing various economic, political and social factors, a feasible hazard mitigation module has been proposed. The module essentially needs government intervention for providing facilities and coordination of various contributors and international organizations for channeling expertise and possible funding. A number of lightning disaster mitigation strategies are proposed and the practicality and limitations of applying them in Uganda are discussed. The proposed module is applicable in many countries with a similar socio-economic atmosphere.

Keywords: *Lightning injuries, Accident(s), Mitigation, Low cost, Developing countries.*

1. Introduction

Lightning related human injuries and fatalities have become an acute problem in many parts of Africa for the last few decades (Mulder *et al.*, 2012; Akello and Ogada, 2006; Blumenthal, 2005; Blumenthal *et al.*, 2012; Dlamini, 2009; Gijben, 2012; Lubasi *et al.*, 2012; Mary and Gomes, 2012, Meel, 2007). Among lightning affected countries in Africa, Uganda has recorded few of the worst incidents in the world history of lightning accidents, including the death of 18 school children and their teacher in 2011 (Mary and Gomes, 2012). Lightning incidents of such large number of multiple victims have been recorded on only a few occasions in South Asia (Gomes *et al.*, 2006; Gomes and Kadir, 2011).

Uganda is a landlocked country with Kenya on the east, South Sudan on the north, Democratic Republic of the Congo in the west, Rwanda in the southwest, and Tanzania in the south. The southern part of Uganda is covered by Lake Victoria, which extends to Kenya and Tanzania as well. Although Uganda is in the equatorial region, the country has a moderate temperature profile due to high elevation. Most parts of the country are on a plateau of altitude around 1000 m. Information issued by the Department of Meteorology, Uganda, states that the country experiences a rainy season from March to May. Light scattered showers are common during November and December. Typically, the country experiences a dry climate from December to February and June to August. The maximum temperature lies between 18 °C to 28 °C, depending on the region.

According to global lightning density maps issued by NASA based on the observations through of the NASA OTD (4/95-3/00) and LIS (1/98-2/3) instruments, Uganda has a lightning density of 10-15 flashes/km^2/year. This value provides a gross underestimation of the level of lightning threats in Uganda. For example Malaysia, a country which falls into 20-25 flashes/km^2/year category, has a smaller number of lightning incidents reported per year.

The recent lightning strike incident assessment carried out in Uganda (Mary and Gomes, 2012), shows that the frequency and intensity of lightning accidents is markedly increasing. Causes of the increase in lightning accidents are not well known. However, one possible reason may be the increase in environmentally unfriendly activities resulting from rapid population growth. People are more exposed to lightning because of the nature of their work. Immediate and appropriate actions are needed to keep the situation from becoming catastrophic, especially that the masses of people that are vulnerable have almost zero awareness of how to protect themselves against this natural calamity.

This short communication is an attempt to formulate a lightning safety module at national level to minimize the adverse effects of lightning despite the limited resources available in developing countries. The module may be applicable not only in Uganda but in many other countries with similar socio-economic environment.

2. Methodology

Data on lightning incidents for the year 2012 were collected from all parts of Uganda. The sources of information were news agencies, personal communications and site visits. The validity of information was cross-checked from multiple sources. Other data and information discussed in the paper was collected from previously published articles and personal communication.

Information regarding the safety modules applied in other countries was collected from previously published literature and personal communication with experts involved with relevant projects. Proposed modules for Uganda were developed based on the analysis of such data.

3. Information Analysis

3.1 Lightning Accident Distribution in Uganda

The first and only investigation so far done on lightning related incidents in Uganda was reported by Mary and Gomes (2012). They analyzed lightning data from 2007-2011. A summary of their analysis is given below. The data was classified into three categories of events: injuries, deaths and incidents.

a. Period of the year: Lightning related incidents show a clear high value during the six month period from June-November with June showing a marked maximum in all three categories

b. Time of the day: Classification was done by dividing the day in to four segments; morning (6 am – 12 noon), afternoon (12 noon – 6 pm), evening (6 pm – 12 mid night) and night (12 midnight – 6 am). All three categories of events showed a clear maximum in the afternoon. This observation was valid irrespective of the month of the year.

c. Part of the country: The Northern and Western provinces show higher number of incidents and casualties than other parts of the country.

d. Place of occurrence: Interestingly, the column for "the place of occurrence of lightning accident" peaks in all three categories in the class of "inside permanent buildings".

e. Year of accidents: Year 2011 shows a remarkable increase in the number of deaths, injuries and incidents. Although one catastrophic incident contributed heavily to the upward trend, the whole country observed an increment in lightning accidents in 2011.

Table 6.1: Summary of Lightning Accidents in Uganda, 2012

Description		Incidents	Deaths	Injuries
Time of the day	Morning (6 am- 12 noon)	2	2	13
	Afternoon (12 noon – 6 pm)	22	30	45
	Evening (6 pm – 12 mid night)	9	23	27
	Night (12 mid night – 6 am)	1	2	0
Place of occurrence	Inside permanent structure	11	24	46
	Inside temporary structure	4	6	7
	In open space	14	14	23
	Under a tree	5	11	29
Part of the country	North	8	15	14
	East	5	11	14
	West	11	17	16
	Central	10	12	31

Contd...

Table 6.1–Contd...

Description		Incidents	Deaths	Injuries
Month of the year	January	1	1	1
	February	1	1	4
	March	0	0	0
	April	2	3	0
	May	6	12	16
	June	5	10	25
	July	0	0	0
	August	8	12	6
	September	7	8	5
	October	3	7	8
	November	1	0	10
	December			
Overall figures	Total	34	56	75

Table 6.1 shows a summary of information on lightning accidents in Uganda collected during the first 10 months of 2012. Although it is too early to draw conclusions about the information distribution for year 2012, a glance at the data reveals that patterns observed during the previous five years have continued without much difference. The characteristic high figures of accidents in 2011 continued in year 2012 as well.

An observation worth noting is that, in Uganda, the period of high lightning accidents is not necessarily the high rainfall period. It has been observed that most lightning accidents happen from June to November while the wettest period is from March to May. Similar observations have been made in the South Asian region that has a monsoon climate, where heavy lightning seasons (March-April and October-November) fall in between monsoons (May-September and December-February) that bring torrential rain (Gomes *et al.*, 2006; Gomes and Kadir, 2011, Lal and Pawar, 2009). A high lightning density during march-April and October-November is observed even in countries with a non-monsoonal climate, such as Saudi Arabia (Shwehdi, 2005). This information also shows that lightning seasons are not uniform even in tropical countries. The data collected also shows that most lightning accidents in Uganda happened in the afternoon. In most of the tropical and subtropical climates, where lightning is brought by convective clouds (Cumulonimbus) such observations are common; *e.g.* lightning accidents in South Asia (Gomes *et al.*, 2006), Singapore (Chao *et al.*, 1981), Australia (Prentice, 1972, Kuleshov and Jayaratne, 2004).

Lightning density is not uniformly distributed in a given country due to geographic and topographic variations. Similarly, lightning accidents may also not follow the lightning density patterns due to non-uniform concentration of population and their socio-economic behavior patterns. Hence, it is not very appropriate to deduce lightning density maps by countrywide isokeraunic data and then to make predictions on lightning accident probabilities based on the derived information (Gomes and Kadir, 2011). Against this backdrop, the real-situation data on region-wise lightning accidents presented in this study and in Mary and Gomes (2012) play a vital role in developing a safety module for Uganda.

In many previous studies it was revealed that lightning accidents rarely happened while the victims were inside permanent structures (Chao *et al.*, 1981, Gomes *et al.*, 2006, Holle, 2009, 2010, Kithil and Rakov, 2001). In contrast to this observation, statistics in Uganda show that the majority of accidents occurred while the victims were in permanent structures. However, close inspection of situations reported in this study as well as those in Mary and Gomes (2012) reveals that almost all those "permanent structures" were thatched/tin roofed, clay/cement/brick walled houses that could not be considered to be sturdily built buildings. Therefore, it is highly important that in the case of developing safety modules, carefully selected wording should be used in defining safety structures.

Figure 6.1: Example of the Houses that are Mainly Affected by Lightning.

3.2 Lightning Safety Modules in USA

Many lightning safety modules have been applied in USA so far and there are some evaluations of their succes (Katie *et al.*, 2000; Holle *et al.*, 1995, 1999, 2005; Ashley and Gilson, 2009; Kithil and Rakov, 2001; Holle, R. E. López, 2003; Cooper and Holle, 2010; Roeder and Jensenius, 2012; Curran *et al.*, 2000).

Lightning awareness campaigns in USA were started on a large scale in the 1980s by government and voluntary institutions, as well as several enthusiastic individuals from different spheres, including medical, engineering, education, recreational backgrounds. In the beginning most of the knowledge dissemination was done through research papers, white papers, newspaper articles etc. However, by the late 1980s, many safety promoters understood that such an academic approach could not reach the masses. Therefore, they thought of developing new modules which were attractive and practically feasible.

School Module

Awareness programmes were organized by various government and non-governmental organizations through which the following activities were conducted:

a. Short seminars and talks by experts in collaboration with various school clubs and societies;

b. Demonstrations and seminars at scouts' and girl guides' camps;

c. Lightning safety related quiz competitions among school children;

d. Web-based lightning safety games and slogans, *e.g.* When lightning roars, go indoors;

e. Distribution and/or displaying of visually effective objects with lightning safety messages: stickers, caps, T-shirts, calendars, notepads, key tags, souvenirs, booklets, pamphlets, banners, signboards etc.

Community Module

Awareness programmes were launched for the benefit of various adult communities. The following activities were conducted:

a. Seminars on lightning safety at various levels of society (at both residential, community and workplace environments);

b. Newspaper articles and public speeches in audio-visual media;

c. Popularizing annual lightning safety week;

d. Public access to online data issued by lightning detection systems in a nation-wide operation;

e. Web-based knowledge dissemination (recently through social networking as well).

Statistics presented by Lopez and Holle (1998, 1996) show that there is a remarkable reduction in lightning casualties in USA over the last century, irrespective of rapid population growth, large scale conversion of bare lands into human settlements and large scale outdoor activities. This is a strong indication that safety modules launched in USA have given rise to fruitful outcomes.

3.3 Lightning Safety Modules in South Asia

In contrast to the USA, South Asia is characterized by a high population density, high lightning occurrence density, high poverty and low literacy levels (except Sri Lanka which enjoys literacy levels above 90 per cent). A majority of the population are employed on the basis of a daily wage, which makes it difficult for them to stay away from work even during thunderstorm. In Bangladesh and eastern parts of

India people's main income is inland fisheries. In many parts of India, Pakistan and Sri Lanka the economy is dominated by agriculture, where the majority of the population are involved in daily work in open fields, predominantly in paddy cultivation.

From 2000 to 2010 a number of lightning safety promotional programmes have been conducted in India, Sri Lanka, Bangladesh, Bhutan and Pakistan by a few organizations and individuals. These programmes were somewhat different to USA modules, as the status of public perception and the level of awareness are different in the two regions.

Programmes in South Asia can be divided into three modules:

Engineering Module

In Pakistan, Bhutan, Sri Lanka and India engineering training programmes have been conducted for almost seven years by lightning protection experts from Sri Lanka and India, with the financial assistance of both government and private sector. Topics covered in many of these programmes include the following;

    a. Lightning protection of structural systems;

    b. Lightning surge protection of HV/MV/LV systems and equipment;

    c. Lightning safety of human beings and livestock;

    d. Electromagnetic compatibility of equipment and systems;

    e. Lightning risk assessment.

Community Module

In Sri Lanka, India and Bangladesh a number of programmes have been conducted to enhance the awareness of general public in safeguarding their lives and property from possible lightning threats. These programmes are basically in the form of;

    a. Workshops and seminars for teachers, community workers, local authorities and civil administrators at divisional and district levels.

    b. Layman's level Programmes for farmers, fishermen, boatmen, housewives, religious leaders, village headmen etc. Programmes were conducted at village centers, bazaars, hats, science and technology centres, schools, temples and mosques, etc. In Bangladesh many modes of activities such as Jarigan (poems) and folksongs, street dramas, storytelling, preaching by religious leaders and scientific demonstrations were conducted. In Sri Lanka most of the programmes were half a day seminars by experts for gatherings of civil servants, community workers and school teachers.

    c. Distribution of awareness materials: Posters, brochure, flyer, calendars, booklets etc.

    d. Mass advertising: Billboards and warning signs at vulnerable places, newspaper advertisements and articles, Street wall-papers (only in Bangladesh), banners at bazaars etc.

e. Formation of lightning awareness centres or lightning awareness units at already existing science and technology centres.

f. Distribution of a lightning safety video in multi languages in lightning safety centres and individuals involved with lightning safety promotion (only in India).

School module: A number of seminars have been conducted for school students during 2004-2009, predominantly in Sri Lanka. Few seminars were also conducted in Bangladesh. Most of the seminars were characterized by speeches by experts which were sometimes followed by demonstrations on first aid treatments and fire fighting.

So far, there has been no evaluation in Bangladesh, Bhutan, Pakistan and India to assess the success or failure of these modules. As per the information presented in Jayaratne and Gomes (2012), irrespective of these programmes, lightning casualties in Sri Lanka shows an upward trend although surprisingly, the survey reveals that the level of awareness and knowledge on both lightning safety and protection among the general public and engineering community in Sri Lanka is extremely high compared to other nations in the region. Jayaratne and Gomes (2012) attribute the increasing lightning hazards, despite the enhancement of knowledge, to the (ignorance) reluctance of public in adopting safety measures that they are aware of. They claim that it is a cultural drawback to the nation.

Another serious drawback to promoting lightning safety in South Asia as it has been observed by many lightning experts in the region (by personal communication) is the extremely high cost of lightning protection systems recommended by international and national standards (IEC 62305, 2010; NFPA 780, 2008; AS/NZS 1768, 2007; SLS 1261, 2004).s It is recommended by Gomes *et al.* (2012) (low cost paper) the international community must make the price of safety modules affordable to developing countries or develop low cost modules for the successful implementation of lightning protection in the developing region.

4. Discussion

4.1 Challenges in Developing Lightning Safety Module in Uganda

Lightning safety module for Uganda should be formulated after considering the failures and successes of modules discussed above and also considering the socio economic environment in the country. A detailed analysis of public perception is also required before developing a successful safety module.

It should be understood that Uganda is a country having high risk of lighting impacts. (Failure of the public in Uganda to adapt to the situation has been as a result of a combination of factors. These factors are the limited knowledge on lightning behavior and socio-economic status of the vulnerable people. Lightning impacts such as injuries and deaths, psychological effects, loss of property and livestock, loss of biodiversity in terms of plants and animals, electric power fluctuations and many others can make sustainability difficult.) The passage in red brackets is not clear. What are you trying to say? Identification and selection of actions to mitigate lightning threats is an uphill battle due to financial constraints. These activities cover a wide range issues, from necessity to sensitize people about lightning characteristics to the development of protection systems for structures and systems that require maximum protection. These strategies will need to be technically feasible and cost-effective if they are to be practically adopted by the public as noted by Gomes *et al.* (2012).

Developing a national action-plan to reduce lightning risk will be a challenging task as causes of the drastic increment of lightning accidents in the country may range from global climate change to national scale issues such as environmental degradation resulting from increased human demand for natural resources. To get the solution to these demands, Uganda's cooperation with international organizations and experts may play a vital role. It is therefore necessary to cooperate through regional and international mechanisms to tackle issues of lightning threats.

In formulating a national action-plan to mitigate the lightning threat which is the interest of this paper, strategies envisaged should be flexible enough to capture various diversities of the nation and specific trends of lightning accidents. Several of these points are discussed below.

a. Uganda is a multi-ethnic and multilingual country. There are over forty indigenous languages that can be classified into four main clusters; *i.e.* Bantu, Nilotic, Kuliak and Central Sudanic. International languages such as English, and regional languages such as Swahili, are also used at government and official levels. Ugandan Sign Language is also popular in the rural areas. Hence the safety module developer will face a big challenge in catering to the entire population of the country. This situation is somewhat similar to India and much different from other South Asian countries and USA.

b. Majority of the population (over 95 per cent) are either Christians (over 85 per cent) or Muslims (over 10 per cent). Hence modules can be easily developed considering sensitivities of the two faiths.

c. Population density in Uganda is around 136 people per km^{-2} as reported in 2011. Compared with USA (34 people per km^{-2} in 2012) the number is quite high. However compared to Bangladesh (1034 people per km^{-2} in 2012), India (382 people per km^{-2} in 2011), Sri Lanka (309 people per km^{-2} in 2012) and Pakistan (228 people per km^{-2} in 2012), Uganda has a much less population density. Furthermore, the population is sparsely distributed hence modules developed should take in to account the difficulties in geographic depth of penetration)

d. As discussed earlier as well a majority of Ugandans live in houses classified as temporary unsafe huts by USA standards. However, there is a sizable urban community who live in sturdily built structures. Most of the rural population work at outdoor sites. However, as per the information given by Mary and Gomes (2012) and this study as well a large portion of lightning accidents happen indoors.

e. A drastic increment of lightning accidents since 2011 should also be taken into account in the development of modules.

In addition to the above discussed points, sharing of responsibilities and uninterrupted operations in executing projects at relevant organizations is also a vital issue for an example, in Uganda responsibilities of issuing lightning safety warning to the public is vested on the meteorological department of Uganda. Thus, most often meteorological department is blamed for un reliable scientific data, yet the government is responsible for funding the installation of modern lightning detectors without which accurate weather casting is almost impossible. However, the public expects the meteorological department to improve its public weather services, despite setbacks.

4.2 Elements of Safety Module

4.2.1 Public Awareness

    a. Need for public awareness: Public awareness on lightning safety plays a critical role in reducing lightning related adverse effects in the country. The need of such awareness has a broad spectrum and most often the importance of such awareness is overlooked due to the lack of understanding of its significance.

        i. To understand lightning safety: Without the public having at least the basic knowledge of scientific concepts of lightning and lightning safety, it is not fruitful to give safety tips. Each safety advice should carry scientific reasoning; otherwise the public may distort the facts by blending them with religious and traditional concepts, leading to disastrous outcomes.

        ii. To use weather predictions meaningfully: There is no point of providing reliable weather forecasting, unless the public have the ability to understand them and act accordingly to prevent weather related hazards.

        iii. To persuade the technical community to adopt proper lightning protection: Availability of lightning protection systems may not be sufficient for the technical community to adopt them for protection. As per the experience in South Asia (Jayartne and Gomes, 2012) even awareness alone is not enough to prevent ignorance? due to cultural trends.

    b. Awareness promotion: The module should ensure that there should be sufficient schemes of awareness promotion that could convey the message to almost everyone in the country, especially those who live in far-remote areas who encounter the highest threat.

    c. Awareness sustenance: In many countries with low level of literacy, it is natural that the level of awareness reduces considerably with time (unless the public perception is repeatedly addressed by various means) unless the awareness promotion is done on an on going basis (experience of authors in South Asia).

    d. Awareness programmes: In contrast to the programmes conducted in developed countries where the programme structures are similar countrywide, in Uganda the programmes should be developed based on accessibility of different social and economic layers. Based on the experience of such programmes in several countries, Table 6.2 has been developed for Uganda. Note that in societies where religious beliefs and societal hierarchy are deeply rooted, it is more fruitful to educate the religious and society leaders first and then convey the safety message to the rest of the community through them

Table 6.2: Summary of Safety Module Contents

Awareness Programme	Programme Features	Target Group
Technical training	Design, installation purchasing or development of LP systems	Engineers, technicians and entrepreneurs at relatively high income ends
School programmes	Expert presentations, video shows, distribution of pamphlets, quiz competitions, classroom lessons, camp-site demonstrations	All schools having at least up to secondary level education
Training of trainers	Expert lectures and demonstrations, site visits, distribution of safety hand-books and demonstration kits etc.	School teachers, community workers, selected personnel from local authorities, health care services, police, emergency services, military etc.
Village programmes	Talks and demonstrations by religious and societal leaders, open-theatre dramas, video shows, posters and banners, distribution of pictures, caps, t-shirts etc. with safety message	Villagers and other low-literacy communities with accessibility to safety promoters
Audio-Visual Programmes	Visual demonstrations and/or audio presentations and expert discussions through TV and radio	All layers of the society, especially those who are not easily reachable by physical means to safety promoters
Programmes through printed media	News paper articles, safety tips in periodicals and magazines and safety books	Social layers with medium and high literacy rate who have accessibility to printed media
Interaction through internet	Video clips, articles, pictures and safety messages, emails, spread of word through social networking	Social layers with high-literacy rate especially those in urban communities

4.2.2 Weather Forecasting

Providing early warning with regard to lightning and thunderstorms is crucial for public safety. The most accurate thunderstorm observation method, available so far, is a country-wide, ground-based lightning detection system. Such system can provide information about the prevailing thunderstorm activities in a given region and forecast the direction of movement of the lightning cells a few hours before the cells discharge. However, as most ground-based lightning detection systems are very expensive at present, there are only a few developing countries that have such systems so far. In this context the following is proposed for Uganda.

a. Existing traditional forecasting methodologies in meteorology sector (both detecting/forecasting and collecting centres) should be upgraded with modern instruments, radars and real time weather monitoring systems.

b. Reaching weather updates to entire public should be ensured by integrating the meteorology services with national and private radio and TV stations.

c. Meteorology department website should be updated on regular basis and the data should be freely accessible to the public.

d. Portable/handheld lightning detectors are much less expensive than large scale systems. Hence such systems should be installed at places of outdoor mass gatherings (such as golf courses, playgrounds, beaches, carnivals, school premises, camping sites, stadiums etc. Such systems can easily be coupled with audio-visual means and implement lightning warning systems at localized areas. Such systems can be implemented in villages-as well.

4.2.3 Low Cost Effective Lightning Protection Systems

Assessment carried out in Uganda indicate that over 70 per cent of the victims were struck inside shelters (Mary and Gomes, 2012).If these shelters had been fitted with lightning protection devices, there's a chance that such incidences wouldn't have happened. In Uganda more than 90 per cent of the population earn less than one US dollar per day. Hence, it is essential to propose low cost lightning structural protection systems (Gomes et al., 2012) and to extend the proposal to countries of similar economic situation, if there is any hope of promoting protection schemes for the general public. Even with low cost systems, unless there is government intervention in providing or at least subsidizing these protection schemes, very few of the society may be able to afford such. Being a developing countries with many priority needs to be fulfilled, it is not possible for the Ugandan government to provide such protection systems to all needy families in the country, hence the intervention of international organizations and funding agencies in this regard is a dire need of the hour.

4.2.4 Lightning Research

Lightning safety and protection of a given country depends on the lightning statics and characteristics of a country. Therefore, it is highly recommended that lightning research in Uganda should be initiated as a pilot project and develop a roadmap to establish a solid research programmein the long-term. Such research project may cover;

a. Characteristics of lightning currents and electromagnetic fields

b. Lightning occurrence distribution of the country

c. Statistics of lightning accidents (human injuries and equipment damage)

d. Human perceptions and societal aspects of lightning

e. Protection devices, techniques and materials

4.2.5 Integration of Lightning Safety with National Development

Lightning safety, especially of a developing country, should be addressed as an integrated part of the national development programme. There are many aspects of lightning safety that are interlinked with other social and environmental concepts. Hence, in the long run lightning safety promotion will be fruitful and sustainable only if the issues are addressed without isolating them from (the entire? Scenario) the national development program

a. Housing for public: One of the key locations of lightning accidents in Uganda is inside so called permanent structures as discussed in section 3. As it is shown in figure-1 they are unsafe huts which can easily be destroyed by lightning. Apart from lightning safety there are many other serious issues that will be faced by occupants of these structures. Hence a national programme is required to upgrade these shelters to more sturdy and long-lasting houses. Low cost lightning protection system can be incorporated into these houses t during the construction stage.

b. Development of national lightning protection standards: It is essential that every country develops its own national standards on lightning protection or adopt international standards with some modifications to suit the local environment (*e.g.* Malaysian LP Standards MS IEC 62305). Formulation of such lightning protection standards should be done by a state administered national institute, in parallel with the development of other standards.

c. Reduction of shading trees for energy needs: Providing adequate, reliable and affordable energy in any country is critical to its overall development because energy is the engine of economic growth and improved lifestyles. The wilderness of the country is diminishing at an alarming rate as 87 per cent of Ugandans depend on wood fuel for cooking and other basic energy needs. Hence for the last few decades a rapid environmental degradation is taking place leaving human settlements more exposure to lightning as they have become the tallest protrusions in the environment. Hence the government needs to look for new and renewable energy sources affordable to the masses to replace fire wood as the main source of energy.

4.3 Action-Plan for Disaster Mitigation

As discussed earlier, in Uganda, the threat of lightning has impacts which need immediate and quick actions to address them. However, the action plan should be carried out by various sectors such as government, non-governmental organizations, academia, voluntary community workers, social leaders and international organizations. The mitigation plan should be long-term in nature, be capable to respond to uncertainties, un-planned events and unanticipated outcomes and be able to get adjusted to suit new information. Irrespective of other stake holders in the mitigation programme, government should either take the leadership or provide total patronage and blessings, to make the endeavor successful. Apart from funding the projects and direct intervention in seeking international funding resources, the government should ensure the smooth coordination and sharing of responsibility among the institutions and stakeholders involved.

Mitigation plan should constitute two essential parts; *i.e.* technical plan and policy plan.

    a. Technical plan: The action plan should take into account the patterns of lightning accidents in the country which have been discussed in section 3. Issues should be addressed in short-term and long term programmes.

       In the short-term:

        i. Comprehensive awareness promotion

        ii. Improvement of weather data accessibility to public

        iii. Provision for low cost effective lightning protectors

        iv. Channeling international expertise

       In the long term:

        i. Accelerated and coordinated research programmes on lightning

        ii. Development of advanced technologies in weather forecasting

        iii. Review planning in relevant fields

        iv. Encourage beneficial behavioral and structural changes in the society

        v. Expand weather observation and monitoring

    b. Policy plan: Several policies and policy instruments are necessary to tackle issues of lightning threat. They can be grouped as economic, administrative, planning, and information. There are differences between them and they address varying audiences. Generally, some criteria in selecting these policies and policy instruments are:

       ☆ Cost effectiveness: Achieving the environmental objectives at the lowest cost.

       ☆ Policy effectiveness: Achieving environmental objectives with the least amount of uncertainty.

       ☆ Dynamic efficiency: Achieving environmental objectives with technological improvements over time.

Economic policies and instruments: Economic policies such as enforcing the need of obtaining license for cutting trees come under this category. Such enforcement encourages reduction of adverse effects due to human action. The government can enforce a law to install mandatory lightning protection to structures that offer occupancy to public (sport stadiums, theaters, large super markets, entertainment parks, art galleries etc.). The state can also direct communication tower owners to provide lightning surge protection and grounding systems to buildings in the neighbourhood. In both cases heavy penalties could be imposed for those who violate the directives.

Administrative policies and instruments: Government initiatives and interventions are most often useful in mitigation strategies. These can be in the form of regulations and standards, administrative guidance on projects and programmes or legal directives. There are several drawbacks in practicing these policies and policy instruments. Compliance can prove to be a major barrier in implementing interdisciplinary directives. Regular dialogue between different stakeholders and establishing mutual recognition and respect increases compliance. The administrative costs to execute some of these measures can be very high for poor countries and so render them ineffective. Poor capacity for enforcement in most countries reduces the importance of these measures (*e.g.* checking for the installation of specified lightning protection systems at all buildings which need mandatory protection).

National planning policies and instruments: Land use planning measures can promote or reduce the demand for energy and hence reduce the severity of lightning threat. Government guidance will be necessary to help promote planning policies that are less environmentally damaging. Problems of compliance and enforcement will be barriers to this approach.

Information policies and instruments: Setting organized information organs, increase public awareness, advisory services and very good education and training programmes can form very good policies and instruments to reduce lightning hazard. One of the major policy decisions that the government can take is to include lightning safety and protection into school curricula.

5. Conclusions

Facts and figures on lightning accidents in Uganda reveal that there are clear patterns of lightning accidents in Uganda so that mitigation plans could be developed taking such information into account. There are many challenges in promoting lightning safety in Uganda such as poverty, illiteracy, disorganized and uncoordinated nature of various institutions that directly responsible for lightning hazard mitigation, lack of modern equipment for accurate weather forecasting, lack of policies and standards etc. However, by developing a module for optimizing the usage of available resources most of the barriers can be overcome and fruitful lightning safety programme could be launched. Programmes should be contributed by various stakeholders starting from village level to top international level. Adequate funding, good coordination among experts and organizing institutes, enforcement of necessary policies and upgrading of forecasting and monitoring equipment play key roles in the success of such lightning hazard mitigation programmes.

References

1. Chao, T. C., pakiam, J. E., Chia, J., A study of lightning deaths in Singapore, *Singapore Medical Journal*, Volume 22, No. 3, 1981.

2. Prentice, S. A., Lightning fatalities in Australia, Electrical Engineering transaction of Institution of Engineers, Australia, 8(2), 55-63, 1972.

3. Lal, D.M., Pawar, S.D., 2009. Relationship between rainfall and lightning over central Indian region in monsoon and pre-monsoon seasons. *Atmos. Res.* 91 (2–4), 402–410.

4. Kuleshov, Y., Jayaratne, E.R., 2004. Estimates of lightning ground flash density in Australia and its relationship to thunder-days. *Aust. Meteorol. Mag.* 53, 189–196.

5. Shwehdi M. H., Thunderstorm distribution and frequency in Saudi Arabia, *J. Geophys. Eng.* 2 (2005) 252–267.

6. López, R.E., and R.L. Holle, "Changes in the number of lightning deaths in the United States during the twentieth century", *Journal of Climate*, 11, 2070-2077, 1998.

7. López, R.E., Holle, R.L., Fluctuations of lightning casualties in the United States: 1959–1990. J. *Climate. 9*, 608–615, 1996.

8. Holle RL, Lo´pez RE, Howard KW, Vavrek J, Allsopp J. Safety in the presence of lightning. *Semin Neurol.* 1995;15:375–380.

9. Lo´pez RE, Holle RL, Heitkamp TA, Boyson M, Cherington M, Langford K. The underreporting of lightning injuries and deaths in Colorado. *Bull Am Meteorol Soc.* 1993;74:2171–2178.

10. Katie M. Walsh, Brian Bennett, Mary Ann Cooper, Ronald L. Holle, Richard Kithil, Raul E. Lo´pez, National Athletic Trainers' Association Position Statement: Lightning Safety for Athletics and Recreation, *Journal of Athletic Training* 2000; 35(4): 471–477.

11. W. Ashley and C. Gilson, "A reassessment of U.S. lightning mortality", *Bull. American Meteorological Soc.*, vol. 90, pp. 1501-1518, 2009.

12. W. P. Roeder and J. Jensenius, "A new high-quality lightning fatality database for lightning safety education," Preprints 4th *Intl. Lightning Meteorology Conf., Vaisala*, Broomfield, CO, April 2012.

13. E. B. Curran, R. L. Holle, and R. E. López, "Lightning casualties and damages in the United States from 1959 to 1994," *J. Climate*, vol. 13, pp. 3448-3464, October 2000.

14. M. A. Cooper, and R. L. Holle, "Mechanisms of lightning injury should affect lightning safety messages," Preprints 3rd Intl. Lightning Meteorology Conf., Vaisala, Orlando, FL, April 2010.

15. R. L. Holle, R. E. López, and C. Zimmermann, "Updated recommendations for lightning safety-1998," *Bull. American Meteorological Soc.*, vol. 80, pp. 2035-2041, 1999.

16. R. L. Holle, R. E. López, and B. C. Navarro, "Deaths, injuries, and damages from lightning in the United States in the 1890s in comparison with the 1990s," *J. Applied Meteorology*, vol. 44, 1563-1573, 2005.

17. R. L. Holle, and R. E. López, "A comparison of current lightning death rates in the U.S. with other locations and times," *Intl. Conf. Lightning and Static Electricity, Royal Aeronautical Soc.*, paper 103-34 KMS, Blackpool, England, 2003.

18. M. B. Mulder., L. Msalu, T. Caro, and J. Salerno, "Remarkable rates of lightning Strike mortality in Malawi," PLoS One, vol. 7:1, doi: 10.1371/journal.pone.0029281, January 2012.

19. R. L. Holle, "Lightning-caused deaths and injuries in and near buildings," *Proc. Intl. Conf. Lightning and Static Electricity*, paper GME-1, Pittsfield, MA, September 2009.

20. R. L. Holle, "Lightning-caused casualties in and near dwellings and other buildings," Preprints 3rd Intl. Lightning Meteorology Conf., Vaisala, Orlando, FL, April 2010.

21. R. Kithil, and V. Rakov, "Small shelters and safety from lightning," *Proc. Intl. Conf. Lightning and Static Electricity, Soc. of Automotive Engineers*, 2001-01-2896, Seattle, WA, September 2001.

22. SLS 1261: 2004, Sri Lanka Standards on Lightning Protection, 2004.

23. IEC 62305 1-4, Ed-01:2010, Protection against lightning, 2010.

24. NFPA 780:2008, American Standards: The Installation of Lightning Protection Systems, 2008.

25. AS/NZS 1768:2007, Australia/New Zealand Standards: Lightning protection, 2007.

Chapter 7

Kasese and Kisoro Non-Directional Beacon (NDB-436) Lightning Protection

Barongo Ronny and Macho David

Civil Aviation Authority
P.O. Box 5536, Kampala Uganda
E-mail: rbarongo@caa.co.ug; rnbarongo@yahoo.com

ABSTRACT

The NDB-436 antenna mast radiates electromagnetic waves. Being a tall structure of about 20m, the antenna mast is prone to lightning strikes. Lightning can cause damage to all or part of the contents of non-directional beacon (NDB) system, especially to electrical and electronic system. To reduce lightning damage, lightning protection measures which include the use of electrostatic protection boards and a static spark gap were put in place for Kasese and Kisoro Non-directional beacons.

Keywords: Electromagnetic waves, Lightning strikes, Electrostatic protection, Antenna tuning unit, Radiating mast, Spark gap, Ground resistance.

Introduction

A non-directional (radio) beacon (NDB) is a radio transmitter at a known location, used as an aviation navigational aid. The NDB enables radio navigation and instrument flight rules, taking away the need for visual landmarks. Therefore, navigation at night and at high altitude is possible.

The NDB-436 uses a mast radiator as its antenna to radiate electromagnetic waves. A mast radiator is a radio tower in which the whole structure itself functions as an antenna. This design is used for the Kasese and Kisoro NDB transmitting antennas operating in LF band (273KHz for Kasese NDB and 258 KHz for Kisoro NDB). The metal mast is electrically connected to the transmitter. At its base, the mast is mounted on a thick ceramic insulator, which has the compressive strength to support the tower's weight and the dielectric strength to withstand the high RF voltage applied by the transmitter. The RF power to drive the antenna is supplied by an antenna tuning unit (ATU), housed in a small metallic box at the base of the mast, and the cable supplying the current is simply bolted to the tower. The transmitter is located in a building near the mast, which supplies RF power to the ATU via a transmission line.

Materials and Methods

Lightning frequently hits tall structures and the NDB radiating mast is no exception. On hitting the structure, lightning finds the easiest way to get to the ground. Lightning discharges can be hazardous to people and equipment connected to the radiating mast. Lightning can cause damage to all or part of the contents of NDB system, especially to the electrical and electronic system. Consequential effects of lightning damage may extend to the surroundings of a NDB system. To reduce lightning damage and its consequential effects, lightning protection measures were put in place both for the Kasese and Kisoro non-directional beacons. The idea is to provide a separate preferential path to ground for the lightning current.

The Ground and it's Functions

A ground is a conducting connection, whether intentional or accidental, between an electrical circuit or equipment and the earth, or to some conducting body that serves in place of the earth. Grounding is at two levels: earth grounding and equipment grounding. Earth grounding is an intentional connection from a circuit conductor, usually the neutral, to a ground electrode placed in the earth. Equipment grounding ensures that operating equipment within a structure is properly grounded. These two grounding systems are required to be kept separate except for a single connection between the two systems. This prevents differences in voltage potential from a possible flashover from lightning strikes. The purpose of a ground besides the protection of people, plants and equipment is to provide a safe path for the dissipation of fault currents, lightning strikes and static discharges.

A Good Ground Resistance Value

Ideally a ground should be of zero ohms resistance. However, due to challenges in obtaining zero ohms earth resistance, the goal is to achieve the lowest ground resistance value possible that makes sense economically and physically. According to NDB technical manual, the whole NDB system should present a ground resistance lower than 4 ohms, in any season or weather condition.

NDB Antenna

An efficient antenna for a Non-Directional Radio beacon would require an effective height of between 600 and 220ft, depending upon the operating frequency in the range of 190 to 535kHz. This can only be achieved by using an extremely tall

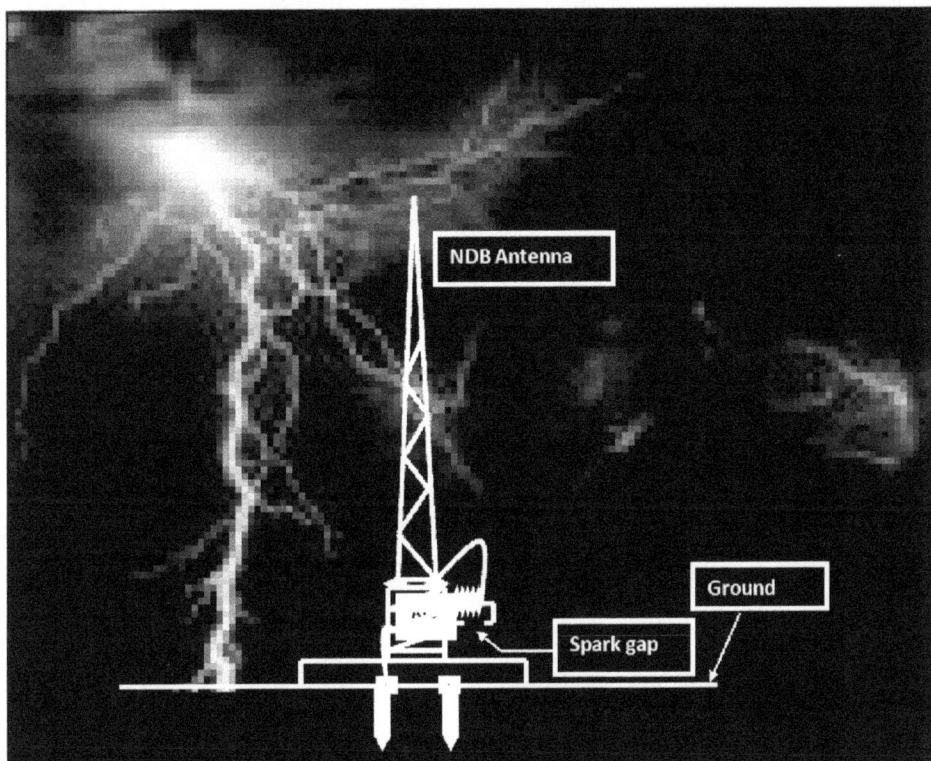

Figure 7.1: NDB Antenna.

structure. That is neither economical nor practical. The effective heights of antennas in common use varies between 20 to 100ft. Both the Kasese and Kisoro NDB antennas are about 65ft (20meters) high.

The spark gap shown in Figure 7.1 is used to protect the NDB equipment from high voltage transients resulting from lightning striking the antenna mast. It consists of an arrangement of two conducting electrodes separated by an air gap, designed to allow an electric spark to pass between the conductors when lightning strikes the antenna mast.

The equivalent circuit of the NDB antenna is shown in Figure 7.2, where:

- ✮ C is equivalent antenna capacitance
- ✮ R_a the radiation resistance of the antenna
- ✮ R_g is the equivalent series loss resistance of the ground plane

When the circuit in Figure 7.2 is tuned to resonance, the input impedance of the antenna system at the carrier frequency is purely resistive.

One of the electrodes that constitute the spark gap taps the connection between the tuning coil (an inductor) and the radiation mast, while the other electrode is connected to the earth.

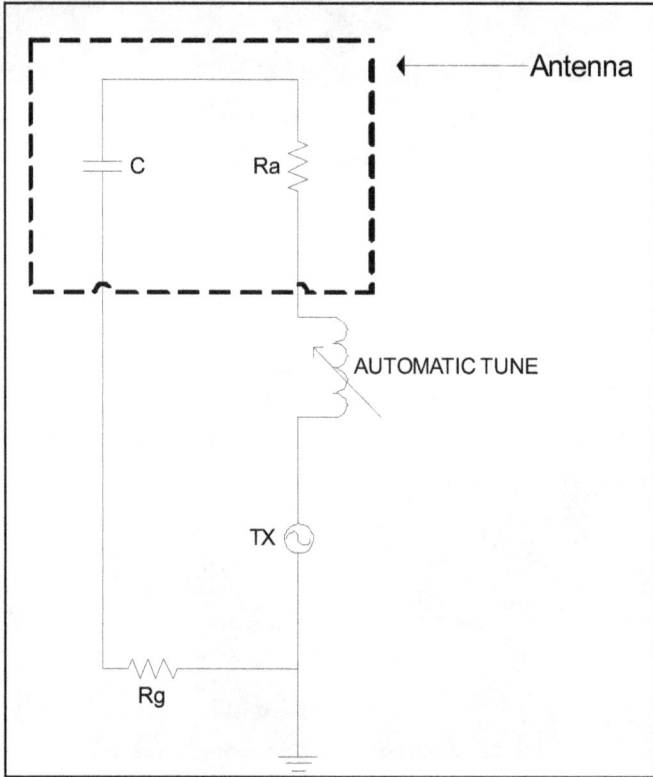

Figure 7.2: The Equivalent Circuit of the NDB Antenna.

The voltage across the terminals of an inductor depends on how quickly the antenna current changes over time. $V_L = L\, di/dt$.

To calculate the voltage across an inductor, the formula is:

$$v = L\frac{dI}{dt} \tag{1}$$

On average, the root mean square value of the antenna current is 3.15 A.

Thus, the current going through the inductor is;

$$I = 3.15(\sqrt{2}[()\sin)](2\pi ft) \tag{2}$$

The voltage across the inductor is;

$$v = L\frac{dI}{dt}$$

$$v = L\,d/dt\{3.15(\sqrt{2})\sin(2\pi ft))]$$

$$v = L(3.15)(\sqrt{2})(2\pi ft)\cos(2\pi ft) \qquad (3)$$

The root mean square value of the voltage across the inductor is therefore;

$$v_{r.m.s} = \frac{L(3.15)(\sqrt{2})(2\pi f)}{\sqrt{2}} = L(3.15)(2\pi f) \qquad (4)$$

To calculate the inductance of the tuning coil (at resonance), we use the following equation;

$$\frac{1}{2\pi fC} = 2\pi fL$$

$$L = \frac{1}{(2\pi f)^2 C} \qquad (5)$$

Using an average value of 400pF as the total capacitance of the antenna, we have:

For Kasese NDB,

$$L_{Kasese} = \frac{1}{(2\pi + 273000)^2 \times 400 \times 10^{-9}} = 850\mu H$$

For Kisoro NDB,

$$L_{Kisoro} = \frac{1}{(2\pi + 258000)^2 \times 400 \times 10^{-9}} = 951\mu H$$

The root mean square value of the voltage across the inductor for Kasese non-directional beacon is;

$$v_{r.m.s} = L_{Kasese}(3.15)(2\pi f)$$

$$v_{r.m.s} = 850 \times 10^{-6} \times 3.15 \times 2\pi f \times 273000 = 4.5kV \qquad (6)$$

The magnitude of the voltage across the inductor is to large compared to the output voltage of the transmitter (typically, less than 100 v). Therefore, it is reasonable to take the voltage drop across the inductor as the same as the voltage across the spark gap.

$$v_{Spark\ gap} = 4.6kV \qquad (7)$$

The root mean square value of the voltage across the spark gap for the Kisoro Non directional beacon is;

$$v_{r.m.s} = L_{Kisoro}(3.15)(2\pi f)$$

$$v_{r.m.s} = 951 \times 10^{-6} \times 3.15 \times 2\pi \times 258000 = 4.9kV \qquad (8)$$

Similarly, the magnitude of the voltage across the inductor is to large compared to the output voltage of the transmitter. Therefore, it is reasonable to take the voltage drop across the inductor as the same as the voltage across the spark gap.

$$v_{Spark\ gap} = 4.9kV \tag{9}$$

The voltages in equations (7) and (9) are below the breakdown voltage of the spark gap. The ionization present at voltages below breakdown is normally too small to affect the engineering application of the spark gap which in this case is overvoltage protection.Further increases in voltage results in additional ionizing processes. Ionization increases rapidly with voltage until ultimately breakdown occurs.

Electrostatic Protection Board

In addition the Spark gap, the use of Electrostatic Protection board protects the electronic components in the NDB transmitter and the ATU.

Figure 7.3: Electrostatic Protection Board.

When a voltage surge occurs, the gas discharge tubes located on the electrostatic protection board shunt high voltages to earth. The gas discharge tubes are self restoring. Therefore, the voltage to which the electronic components of the NDB equipment are exposed to is limited to a level much lower than that of the surge.

The gas discharge tubes provide reliable and effective protection solutions during lightning storms and other electrical disturbances. A gas discharge tube does wear out due to particulates being dislodged from the electrodes during the arcing of the tube. The impact of the arc across the tube is dependent on the energy strike, so the life of the gas discharge tube will be dependent on the impulse applied to it.

If the electronic components of the NDB equipment are damaged by lightning, and the electrostatic protection board isn't damaged, that means there was a bad ground (or no ground) attached to the electrostatic protection board. The electrostatic protection board should fail before a damaging voltage or current can get to the electronic components of the NDB equipment it is protecting.

Results and Discussion

Instances of Lightning Strikes on the Kisoro NDB

Kisoro NDB was installed at Kisoro airfield in 2002. Kisoro is a mountainous rocky and volcanic area known for frequent lightning strikes and thunderstorms. It is about 4200m above sea level compared to 1000 meters above sea level for other parts of the country. Two transmitter units blew within 1 year after installation and were replaced under manufacturer's warranty (each transmitter unit cost approx. US $10,000). In 2004, the antenna capacitive top loading and NDB electronics were hit by lightning. It took about a year to fix the problem and cost the company both civil and electrical works approx. US $ 40,000. At least every year the equipment goes off due to lightning and voltage surges.

Other Air Navigations Systems Affected by Lightning

Soroti DVOR (costing about Euro 400,000) was replaced in 2006 after a lightning strike damaged its transmitters and power supplies.

Entebbe DME' s circulator was destroyed by lightning in 2004 and replaced after 1 year.The VHF communications system was hit by lightning in 2009. Insurance made a compensation of Euro 50,000 for spare parts. The Mother Board of the Glide Slope Entebbe was struck by lightning in 2007. It cost Euro 10,000 and took 1 year to replace it.The VSAT Terminal transmitters were hit in 2010. Remote Control and Monitoring Equipment (RCSE) and its modems were hit by lightning in 2008.The RCSE was upgraded since there were no spares available on the market.

Radio antennas at Kigulya Hill in Masindi District were burnt by lightning in 2009 and replaced after 1 year.

Conclusion

Lightning is a serious danger that should never be underestimated. Risks associated with lightning strikes can be highly mitigated by the proper implementation of a surge suppression and grounding systems.

References

1. Department of Transportation, Federal Aviation Administration Standard "Lightning and surge protection, grounding, bonding and shielding requirements for facilities and electronic equipment" FAA-STD-019e, December 22, 2005.

2. Elvis R. Sverko "Ground Measuring Techniques: Electrode Resistance to Remote Earth and Soil Resistivity" ERICO, Inc. Facility Electrical Protection, U.S.A..

3. FLUKE, "Principles, testing methods and applications - Earth Ground Resistance"2006 Fluke Corporation.

4. Travis Lindsey, "Technical Report - National Electrical grounding research project" The fire protection research foundation, August 2007.

5. Airsys Navigation Systems "NDB-436 Non Directional Beacon" Volume 2, BASE, June 1998.

6. Martin D. Conroy and Paul G. Richard "Deep Earth Grounding vs. Shallow Earth Grounding" Computer Power Corporation Omaha, Nebraska.

7. John Pinks "NDB Antennas" 2003 NAUTEL.

Chapter 8

Are Lightning Injuries Different in Developing Countries?

Mary Ann Cooper

MD Professor Emerita
University of Illinois @ Chicago (retired)
632 Clinton Place, River Forest, IL 60305, USA
E-mail: macooper@uic.edu

ABSTRACT

Multiple reports have reviewed the medical aspects of lightning injury but they are primarily based on cases that are taken from developed countries [1-3]. Often, accounts in developing countries report much more severe injuries than usually occur in developed countries. Insufficient data exists to determine if these reports are exceptions or actually reflect a difference in injury pattern. This paper will examine potential reasons for bias in these reports as well as propose hypotheses that may account for any true differences in the injuries.

Keywords: *Lightning, Lightning injury, Lightning injury prevention, Lighting protection, Burns, Heraunoparalysis, Mechanisms of injury, Fire.*

Introduction

In reading reports from developing countries, it is striking how often the descriptions 'burned beyond recognition' or 'charred' show up. In addition, at least some of the pictures that this author has seen led her to question not only whether injuries are more severe in developing countries but the reasons this could occur.

As a first examination of reasons for increased or different severity of injury, it is reasonable to analyze all aspects of lighting injury from the formation of lightning to the injury reports to the final outcome (Table 8.1).

Table 8.1: Factors that Could Account for the Observation of more Severe Injuries in Developing Countries

☆	Physical characteristics of lightning in developing countries
☆	Mechanisms and distribution of injury mechanisms
☆	Reporter error and bias
☆	Sampling bias
☆	Misinterpretation of severity based on injury complications,delay in treatment, infection, sampling bias, photography

Definition of Developing Countries

Most commonly the criteria for evaluating the degree of development of a country includes measures such as gross domestic product (GDP), the per capita income, level of industrialization, amount of widespread infrastructure and general standard of living. For the purposes of this paper, 'developing country' will be taken to mean countries which, despite some advanced urban areas or new industrialization, have a large proportion of their population employed in labor intensive work, living in ungrounded and often easily destroyed structures, and usually limited access to goods, medical care and education. The majority, but certainly not all, of these countries lie in tropical and subtropical areas of the world.

Documentation of Lightning Injuries in Developing Countries

It has been estimated that lightning causes at least 24,000 deaths and 240,000 injuries annually in the tropical and subtropical areas of the world [4]. Studies have shown that the US has approximately ten injuries for every death [5], so it has become routine to multiply the estimated deaths by ten to provide injuries estimates in the US. However, it is unknown if this 10:1 ratio applies to developing countries. In addition, the true number of people injured and killed by lightning in most developing countries is unknown [4,6].

In the US, only 'direct' or primary lightning injuries are included in the statistics. For instance, those who die in a fire or building collapse caused by lightning strike to a structure would not be classified as a lightning death but listed under burns/fires or building collapse. Particularly for developing countries, at least for the first few years of data collection (as was done in the US as well), it may be desirable to catalog all deaths and injuries precipitated by lightning in order to highlight areas amenable to prevention strategies.

Physical Aspects of Lightning in Developing Countries

It is beyond the expertise of this author to discuss whether there are characteristics of lightning in developing countries that would affect the quality of injuries.

Mechanics of Injuries in Developing Countries

Mechanisms of Injury

There are several direct or primary mechanisms by which lightning can cause injury [1]. The distributions of these have been previously reported [7] and are shown in Table 8.2 and Figure 8.1.

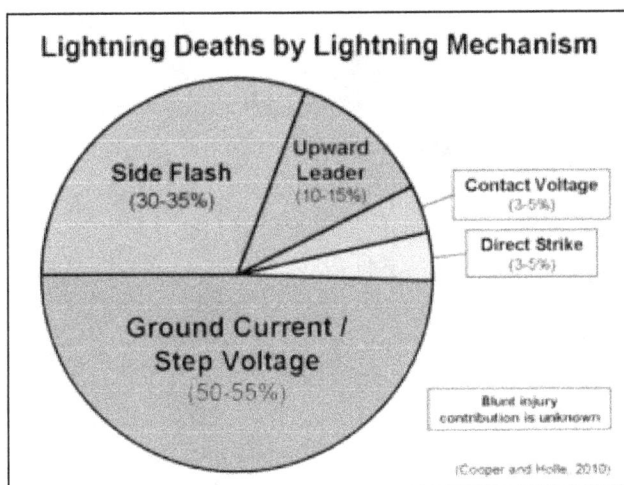

Figure 8.1: Distribution of Lightning Injury Mechanisms in Developed Countries.

Data on the distribution of lightning injuries was taken primarily from newspaper reports in developed countries where housing is substantial, metal vehicles are readily available for shelter from storms, and news reports potentially more likely to involve fewer iterations between the event and the reporter making the report.

Table 8.2: Distribution of Lightning Injuries by Mechanism in Developed Countries [7]

Mechanism	Percentage
Direct Strike	3-5 per cent
Contact Injury	3-5 per cent
Side Splash/Flash	30-35 per cent
Ground Current	50-55 per cent
Upward Streamer	10-15 per cent
Blunt Injury	Unknown

It is unknown if the distribution of mechanisms is different in developing countries where open air schools, homes (Figures 8.2–8.3) and vehicles (Figures 8.4–8.5) are less likely to have a Faraday cage effect, where indoor plumbing and wiring is less available to transmit a contact injury to victims, where trees and other tall objects for side splash to occur may be less common, where labor demands and personal exposure is different, and where other unstudied variables may be major factors (Table 8.3).

**Figure 8.2: Ungrounded Metal Roofed Building.
Photo India, M.A. Cooper.**

**Figure 8.3: Thatched Dwelling and Family Business in India.
Photo M.A. Cooper.**

**Figure 8.4: Open Sided Taxis.
Photo India, M.A. Cooper.**

**Figure 8.5: Open Cart Used for Transporting Goods and People.
Photo India M.A Cooper.**

Table 8.3: Factors that May Affect Injury Mechanism Patterns and Injury Prevention between Developed and Developing Countries

Factor	Develped	Developing
Injury Statistics	Reasonably good	Largely unavailable/unknown
Population	Largely urban	Largely rural
Farm/work	Mechanized	Labor intensive
Transportation	Mostly enclosed metal vehicles	By foot, bicycle or unprotected cart
Housing, schools	Good construction, 'Faraday cage' effect	Open, roofed with metal or flammable thatch
Victims	Primarily outdoor workers or sports participants	Entire classroom or family sleeping insided wellings
Distribution of Injury Mechanisms	Reasonable data from a large number of reports	Different terrain and lack of safer structures maychange distribution
Injury/Death ratio	10:1	Research could include indirect injuries at first

Indirect Injury

In addition to the direct or primary electrical mechanisms of injury where the electrical aspect of lightning is the causative factor, injury may be worsened or the number of injured increased by indirect or secondary effects [8] such as when a lightning struck home collapses around a sleeping family or the thatched roof (Figures 8.6–8.7) catches fire and some members, particularly the younger or older ones, are not mobile enough to escape. Keraunoparalysis [1,9], temporary paralysis of the legs and sometimes arms, often occurs immediately after a lightning injury and may prevent even otherwise healthy adults from escaping (Figure 8.6).

Figure 8.6: Rondavel in South Africa where this Woman's Family Member Died by Fire Photo Courtesy: Ron Holle.

Figure 8.7: Thatched Dwellings in India. Photo M.A. Cooper.

Flame burns often have significantly higher complication and mortality rates in developing countries. While these are not 'technically' direct or primary lightning injuries, cataloging and studying them may lead to specific preventive strategies that could save many additional lives.

Ungrounded, insubstantial dwellings, open schools and the potential for fire increases the population at risk from primarily outdoor workers to both the workers and their families 24 hours/day in lightning prone areas (Table 8.3).

Factors Leading to Errors in Media and other Reports of Injury [7]

Many of the reasons for misreporting of lightning cases in developed countries [7] are likely to apply in developing countries as well.

Lightning injury is almost always a dramatic and unexpected event. Lightning is so sudden and its course so rapid and variable that the human eye cannot record it accurately. Even well trained lightning researchers have been amazed at the differences between their visual and instrument observations and the revelations that have occurred with recent high-speed lightning photography.

Few incidents are investigated by qualified experts and few criteria exist which define the evidence that should be sought for lightning strikes involving personal injury. The vast majority of newspaper and personal reports in developed countries credit direct strikes as the cause, largely in part due to lack of knowledge of other well-accepted mechanisms. In addition, misreporting may occur for many reasons including lack of knowledge of other mechanisms by the victim, reporter, or witnesses, errors in observation and assumptions by eyewitnesses untrained in lightning, victim amnesia, and over-dramatization of the event. None of these may be intentional but do lead to errors in data collection and interpretation, prescription of lightning avoidance tactics, and perception of lightning risk by both the public and many professionals.

Of necessity, media reports are retrospective but, at least in developed countries, are often gathered from witnesses and survivors of lightning strike. When injuries occur in developing countries, particularly in remote areas, the account of a lightning incident may be delayed by several days and go through several people before reaching authorities or newspaper reporters. Each iteration of the story adds another layer where misinterpretations, misinformation and embellishments may creep in. In other areas, reports may be suppressed due to taboos or other beliefs about the causes of lightning injury. Language or dialect differences may also play a role.

The majority of reporters are unlikely to have in depth knowledge of lightning science or lightning injury mechanisms. Even in developed countries, misinformation and myths such as metal attracting lightning or rubber tires or rubber soled shoes 'saving' the person often creep in. The most well-intentioned report can have serious inaccuracies, which, while not dangerous to anyone, nevertheless cloud data collection efforts.

In reading reports from developing countries, it is striking how often the phrases such as 'burned beyond recognition' or 'charred bodies' are used. It is unclear whether these are accurate accounts or a reflection of what the reporter simply expects to be the findings when they have no first-hand information. This also leads to sampling bias and data errors.

Ron Holle, international demographer of lightning injuries states, 'It's really hard to know what to make of these, except there are a lot of fires [10].'

Sampling Bias

Due to the number of reporters, the excessive public demand for instantaneous news, and the need to fill media space in the United States, it is doubtful that any lightning deaths in the last few years have escaped the databases maintained by the National Weather Service [11]. Deaths are generally better reported than injuries [12] since there is no registry or mandatory reporting of lightning injury [5,12] so that the actual number of injuries may be considerably higher than reported. In addition, most survivors do not require a hospital admission and this will affect any database which relies on hospital admission statistics [5].

In many other countries, lightning injury statistics are based on news reports or medical records [4,6]. It is highly likely that many are missed. In addition, at least some sampling bias is present in the overall population of reports. One would expect that the more severe cases and those involving multiple victims would be the ones most likely to make it to the media's attention, especially in areas where information on individual deaths and injuries rarely spreads beyond the small village where the incident occurred. This would overweight data on injury severity and multiplicity collected from media reports.

Potential Reasons for Misinterpretaion of Injury Severity

In developed countries, lightning burns are reported in less than one third of survivors [1,13]. They tend to be superficial and minor, usually requiring little more than topical care and resolving with little or no scarring, similar to what a normal sunburn would do. Only a very tiny number require grafting or more aggressive care similar to high voltage burns.

Table 8.4: Factors that Could Account for Errors in Injury or Burn Severity Interpretation

Factor	Explanation
Sampling error	Only the most dramatic photographed
Secondary burns	Flammable thatch, trapped inside
Delayed care	Infection enlarging or swelling wound and surrounding tissues Home remedies may worsen appearance Infestation Normal burn edema swells tissue surrounding burns 2-5 days after burn
Post mortem changes	
Errors of skin photography	Colors not true or washed out, only two dimensions depicted

However, media reports in developing countries often contain 'charred,' 'burned beyond recognition,' 'blackening of the skin,' or similar phrases. It is unclear how many of these are based on first person accounts or direct evidence and how many may be embellished by reporter assumptions and expectations. It is also quite likely that sampling error, where only the most dramatic wounds are documented, could be a factor. There are other plausible explanations that may influence either the reporting or the appearance of the wound and photographs of them (Table 8.4) that are difficult for anyone who is not experienced with burns or who does not frequently see these types of photos to appreciate.

It is very difficult to capture true skin color and changes when filming close enough to see the wound without special lighting techniques. Anyone who has ever tried to photograph a rash or skin changes, including burns, will almost certainly be disappointed in the photos which, for many reasons, seldom portray the wounds as clearly and dramatically as first hand observation where nearly all of the senses are involved. Nevertheless, pictures of injuries and burns from developing countries that appear in the press or have been circulated between lightning professionals sometimes appear to be much more severe than burns typically observed in developed countries. There are even terms ('tip-toe sign') that are unfamiliar to US professionals that are used to describe these.

It is worth further investigating to find out if there are indeed differences in the wounds observed in developing countries. Delay in care, use of local or native treatments, normal changes that occur in burns as they 'mature', post-mortem changes and other factors may change the appearance of burns by the time they are photographed, treated or observed by investigators (Table 8.4).

It is likely that whatever explanations are discovered for real differences in injury patterns, if they exist, need to be incorporated into injury prevention education [14].

Lightning Injury Prevention

Recommendations that have helped to decrease lightning injury in the United States [1,6,11,15-19] and other developed countries may not be practical or achievable in developing countries for many reasons (Tables 8.3 and 8.5). Factors such as substantial construction, good communication systems, and large numbers of metal vehicles may be unattainable, impractical to expect or unaffordable in developing parts of the world [6,14].

Table 8.5: Factors that may Complicate Injury Prevention Efforts in Tropical and Subtropical Developing Countries

☆ Higher incidence of lightning

☆ Higher exposure due to work practices

☆ Lack of safe structures for housing, work and schoolsincreasing exposure risk to 24/7 for everyone

☆ Multiple languages, long distances, poor access

☆ Communication systems may be poor

☆ Myths and superstitions

Messages used for public education in one country or area may need to be modified to be effective in another country due to cultural, religious, language or employment differences [14]. For instance, recommending that someone seek shelter in a substantial building or metal vehicle is useless when neither of these is available to a seasonal farm worker on the high planes of Kenya.

There are many components to be considered as parts of any injury prevention programme (Table 8.6).

Table 8.6: Selected Components of Lightning Injury Prevention Programmes [14]

Factor	Implementation
Risk recognition	Data collection, government involvement
Raising awareness	Media involvement
Multidisciplinary team	Professionals, parents, media
Tailoring and delivering the message to the audience	Street plays, coach training, YouTube, community leader involvement, use of prominent spokespersons/sports figures
Changes in building codes, public safety rules [20-23]	Development of cheap, effective, easy to maintain protection methods for homes, businesses, schools, workers

Conclusion

It is important to realize that most lightning injury literature has come from reports and data from developed countries. The distribution of mechanisms of injury and, indeed, the severity and range of injuries may be different in developing countries but has not been investigated in an organized fashion.

It is essential that those interested in injury prevention, including experts in lightning protection, work together to collect data, formulate injury prevention programmes, and design innovative, affordable and effective protection systems that can be used in developing countries.

Acknowledgement

I would like to thank the lightning specialists who have patiently tolerated my questions and taught me over the past three decades including E. Philip Krider, Martin Uman, Ken Cummins, Vlad Rakov, Michael Cherington, John Jensenius, William Roeder and others too numerous to mention. There is no way to adequately thank my good friends Ron Holle and Chris Andrews who have been sounding boards, editors, co-authors and colleagues for over twenty years. Ryan Blumenthal, Zainal Kadir, Chandima Gomes, Shriram Sharma, Estelle Trengrove, Lubasi Foster Chileshe, Richard Tushemereirwe and many others have been instrumental in expanding my view of lightning injuries beyond the US.

References

1. M.A. Cooper, C.J. Andrews, and R.L. Holle, "Lightning Injuries" in Wilderness Medicine, 5th ed, P. Auerbach, Ed: 5. St. Louis, CVMosby. 2006, pp. 67-108. Available at: http://www.uic.edu/labs/lightninginjury/LtnInjuries.pdf.

2. M.A. Cooper, "Are the medical aspects of lightning injury different in developing countries?" Unpublished.

3. M.A. Cooper, "Medical aspects of lightning injury," in "Lightning Protection", Shriram Sharma, Ed. Daya Publishers, New Delhi, India, 2012 (in press).

4. R.L. Holle, "Annual rates of lightning fatalities by country," International Lightning Detection Conference, Tucson, Arizona, 2008. Available at http://www.vaisala.com/Vaisala per cent 20Documents/Scientific per cent 20papers/Annual_rates_of_lightning_fatalities_by_country.pdf

5. M. Cherington, J. Walker, M. Boyson, R. Glancy, H. Hedegaard, and S. Clark, "Closing the gap on the actual numbers of lightning casualties and deaths," *Preprints, 11th Conference on Applied Climatology*, Dallas, January 10-15, Boston, American Meteorological Society, 1999.

6. M.Z.A. Kadir, M.A. Cooper, and C Gomes, "An overview of the global statistics on lightning fatalities," *International Conference on Lightning Protection*, Cagliari, Italy, 2010.

7. M.A. Cooper, R.L. Holle, and C.J. Andrews, "Distribution of lightning injury mechanisms," *International Conference on Lightning Protection*, Cagliari, Italy, 2010.

8. R.L. Holle, "Lightning–caused deaths and injuries in and near dwellings and other buildings," *4th Conference on the Meteorological Applications of Lightning Data 2009, Phoenix*, Arizona, American Meteorological Society.

9. M.A. Cooper, "Lightning injuries: prognostic signs for death," *Ann Emerg Med*, vol. 9, pp. 134-139, 1980.

10. R.L. Holle, personal communication.

11. Lightning safety awareness (LSA) website: www.lightningsafety.noaa.gov

12. R.E. Lopez, R.L. Holle, T.A. Heitkamp, M. Boyson, M. Cherington and K. Langford, "The underreporting of lightning injuries and deaths in Colorado," *Bull. Amer. Meteor. Soc.* 74, 2171-2178, 1993.

13. C.A Pfortmueller, Y. Yikun, M. Haberkern, E. Wuest, H. Zimmermann, and A.K. Exadaktylos, "Injuries, sequelae, and treatment of lightning-induced injuries: 10 years of experience at a swiss trauma center," Emerg Med Int 2012.

14. M.A. Cooper and M.Z.A. Kadir, "Lightning Injury Continues to be a Public Health Threat Internationally," International Lightning Meteorology Conference. 2010 Available at http://www.vaisala.com/Vaisala per cent 20Documents/Scientific per cent 20papers/5.Cooper, per cent 20Zainal.pdf

15. M.A. Cooper and R.L. Holle, "Mechanisms of lightning injury should affect lightning safety messages," International Lightning Meteorology Conference, Orlando, Florida, USA, April 21-22 2010 Available at http://www.vaisala.com/Vaisala per cent 20Documents/Scientific per cent 20papers/11.Cooper, per cent 20Holle.pdf

16. American Meteorological Society, "Lightning safety awareness [AMS Statement and background from AMS Council]." *Bulletin of the American Meteorological Society*, 2003, vol 84, pp. 260-261.

17. Zimmermann, M.A. Cooper, and R. Holle, "Lightning safety guidelines," *Annals of Emergency Medicine*, 2002, vol 39, pp. 660-665.

18. K.M. Walsh, B. Bennett, M.A. Cooper, R. Holle, R. Kithil, and R. López, "National Athletic Trainers' Association position statement: Lightning safety for athletics and recreation," *Journal of Athletic Training*, 2000, vol 35, pp. 471-477. (Currently in revision).

19. B.L. Bennett, R.L. Holle, and M.A. Cooper, "Lightning safety," 2007-07 NCAA Sports Medicine Handbook, 18th Edition, D. Clossner, Editor, National Collegiate Athletic Association, Indianapolis, Indiana, 12-14.

20. Energy Commission of Malaysia. Circular No. 3/2011: Declaration of Lightning Protection System Installation for Buildings. ST(IP/JKE)PE/16/04/1(3).

21. British Standards Institute. BS EN 62305-2. Protection against lighting: Part 2- Risk management.

22. National Fire Protection Association, Standard for the Installation of Lightning Protection Systems, NFPA 780, 2008.

23. International Electrotechnical Commission. IEC 62305-1. Protection of Structures Against Lightning: Part 1- General Principles.

Chapter 9

Country Status Paper on Lightning Protection, Zambia

Foster Chileshe Lubasi (Ms)

National Institute for Scientific and Industrial Research,
Zambia
E-mail: fclubasi@yahoo.com

ABSTRACT

Situated in southern Africa, Zambia experiences tropical climatic conditions modified by a sub-tropical climate. The country's rainy season is from the end of October to early April. The distinction between rainy and dry seasons is marked, with no rain at all falling in June, July and August. The rains are characterized by thunderstorms, quite often severe, with much lightning and sometimes hail [1].

Research in lightning protection is still in its infancy. A few case studies were done on lightning incidents around the country that occurred during the 2010/2011 and 2011/2012 rainy seasons and information collected is in the process of being interpreted scientifically with the help of University Putra Malaysia, UPM. The long-term objective is to systematically study the effects of lightning on the social and economic development of the country, to propose appropriate protective devices and systems for electronic equipment, people and livestock and to help government to propose legislation on lightning protection. The few case studies conducted revealed that, in the 2011/2012 rainy season alone the country lost at least 19 people in various provinces, 18 cattle and electronic equipment, some of it very critical, like the entire local area network for the meteorological department of Zambia, the internet system for the Zambia police in a town in Northern province and the radio network for the Zambia Police in one of the villages in Western province. It is hoped that this project will eventually get the attention and support of the relevant authorities.

Keywords: Lightning, Lightning strike, Earthing, Grounding, Lightning incidents.

Introduction

Situated at approximately 15° S and 30° E [2], Zambia is a land locked country in the southern part of the African continent. The country is surrounded by eight countries: Democratic Republic of Congo, Tanzania, Malawi, Mozambique, Zimbabwe, Botswana, Namibia[3]. Figure 9.1 is the map of Africa with Zambia highlighted.

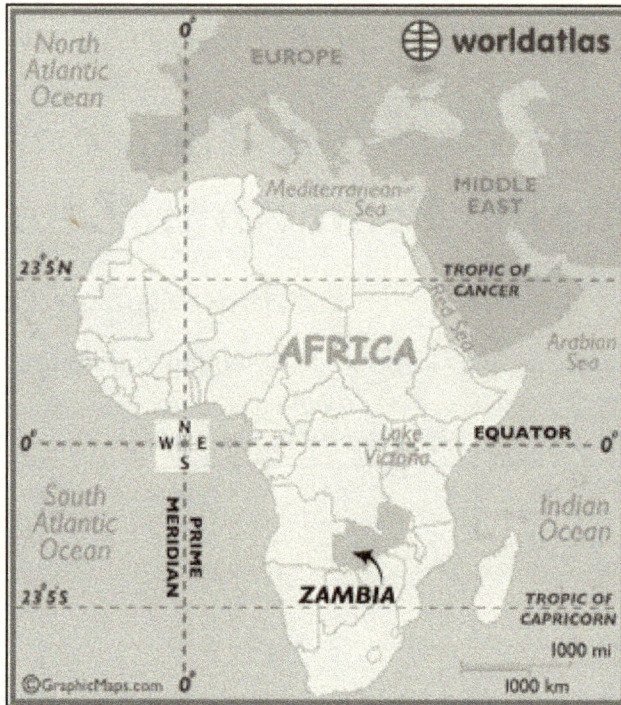

Figure 9.1: Position of Zambia in Africa [4].

Zambia's lightning flashes per square kilometre have been documented as being between 10 to 20 flashes per kilometre squared based on National Aerospace and Space Administration, NASA satellite photography, shown on the map in Figure 9.2.

Zambia has not yet done it's own research to verify the NASA satellite images. However suffice to state that every rainy season Zambia experiences a high level of lightning activity that is often fatal, striking buildings, human beings, livestock and damaging electronic/electrical equipment.

Site visits were carried out by a team of researchers under the Materials Engineering and Technical Services Programme of National Institute for Scientific and Industrial Research, Zambia, during the period from January to June, 2012. The cases investigated were mainly for the 2011-2012 rainy season and a few for the 2010/2011rainy season lightning incidents. The investigations were carried out under the guidance of the Centre of Excellence on Lightning Protection, the University Putra, Malaysia.

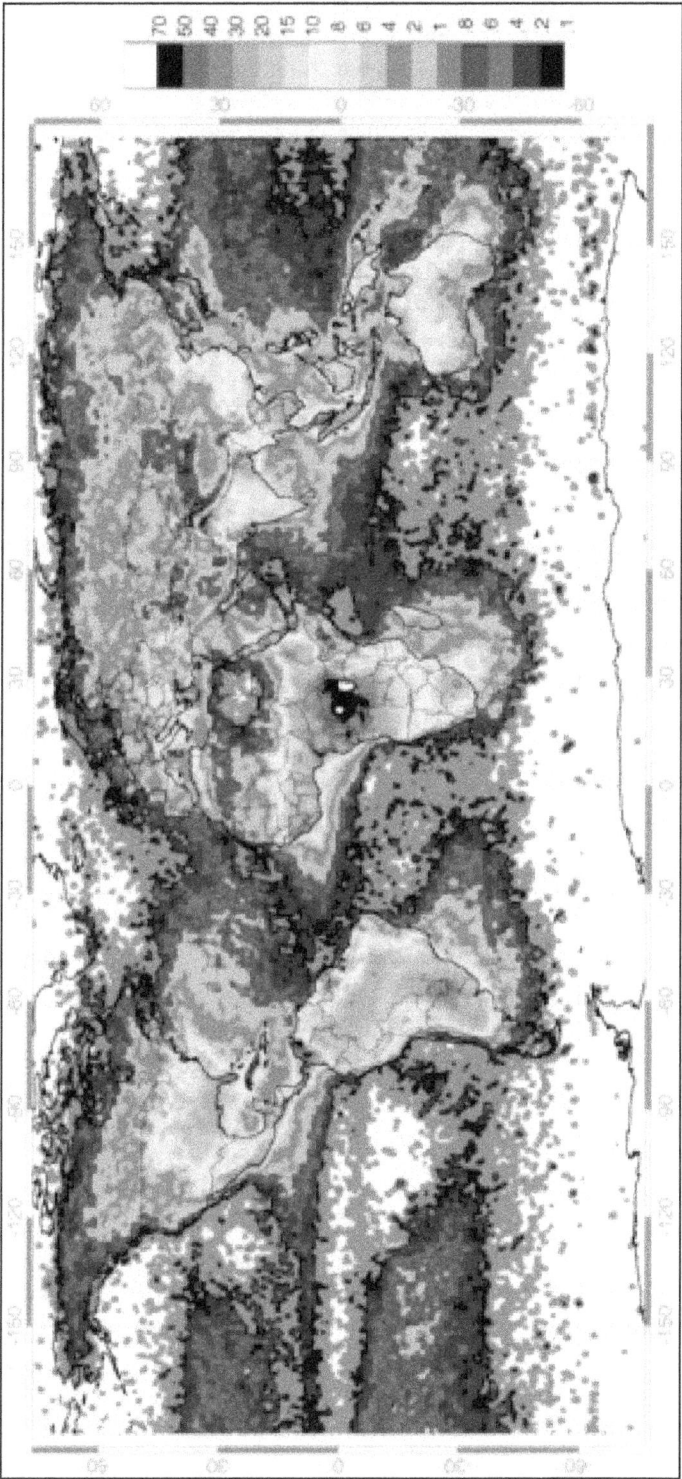

Figure 9.2: Distribution of Worldwide Lightning Strikes (flashes/km²/yr) [5].

The sites for investigation were selected by considering convenience of access, rather than any scientific criteria. Since Zambia is a country of large span with many areas not easy to reach, selecting sites for lightning injury/damage investigation was not an easy task, hence, the research team visited whenever and wherever possible to collect data. Pre-information regarding the incidents was collected either from media reports or by personal correspondence.

During the visits, survivors, eye-witnesses, neighbours and relatives of affected people were interviewed. Photographs of affected buildings, survivors and eye-witnesses were taken where possible. The information gathered has been cross-examined through various other sources such as media reports, records maintained by any authority mainly the police stations or evidence given by a reliable involved third party.

The institution could however not manage to study all the cases brought to the attention of the researchers while in the field due to budget and time constraints. However, the few cases studied in the provinces indicated below is an indicator that lightning strikes kill more people, livestock and damage more equipment than the public is aware. More work is required to establish the actual figures and to also establish the actual characteristics of lightning in Zambia and how it interacts with people, livestock and buildings in order to come up with appropriate measures for protection.

Work Done

As stated above, National Institute for Scientific and Industrial Research conducted some case studies on lightning strikes from various parts of the country. The provinces visited were, Muchinga province in the northern part of the country, Central province, Northern province and Western province. In the provinces visited, only the following areas were visited and only a few selected cases were studied.

Muchinga Province

In Muchinga province the places visited were Nakonde town center, Itindi and Chozi villages.

1. At Nakonde town centre lightning struck a shelter under which people were playing pool and others took shelter from the rain on 10th December, 2011. Several people were affected but only two died.

2. In Itindi village, lightning struck a church building and affected about ten people who were inside. However no one died all of them just fainted but later regained consciousness. A number of them still suffer after effects of the lightning strike in various parts of their bodies. Some complained of experiencing chest pains from time to time, others complained of pains in the legs and others complained of experiencing pain in the shoulders from time to time.

3. In Chozi village, a Zambia Electricity Company pole-site was visited which had been struck by lightning on 2nd April, 2012, leaving the insulators and overhead cables broken and set the corn field through which the cable passed on fire.

Central Province

In Central Province, the towns and villages visited were, Mkushi, Kapiri Mposhi, Kabwe, Chibombo and Mumbwa.

1. In Mkushi, the research team visited Kasansama village were lightning struck a house killing five of the six people who were in the house. These were all members of one family. A little boy estimated to be about three years old survived.

2. Kapiri Mposhi is also a small town in Central province. In Kapiri Mposhi, only one site was visited. The site was a tower belonging to Zambia Railway Systems that was struck by lightning in October 2011. In the process, the armoured cable that supplied power to a remote radiation monitoring system was burnt. Power was only restored to the system in July 2012. After the power was restored, it was discovered that the electronic system had also been affected. At the time of writing this paper, the system was still non-operational. It is not able to detect radiation and, it was not yet known how much it was going to cost to replace the signal processor board. The lightning strike on the Railway Systems of Zambia blew up the cable because the earth bar for the tower was broken. The high lightning current would have most likely gone to ground through the earth bar had it not been broken and would not have damaged the portal monitors.

3. In Kabwe the team visited two (2) sites in the same compound called Katondo. On 1 November 2012, lightning struck the compound and affected quite a wide area. Two boys died in two separate homes about 100m apart. The strike affected an area estimated at 400m in diameter and destroyed

Figure 9.3: Railway Systems of Zambia Tower.

**Figure 9.4: The Classroom and the Tree that was Struck by Lightning.
The building has been abandoned since the incident.**

**Figure 9.5: The Cousin and Sons of the Deceased Stand in the Position where the
Man had been Sitting with his Two Sons when he was Killed by Lightning.**

electronic equipment ranging from Hi Fi systems, television sets and DVD players in various home.

4. In Chibombo, the team visited Nchename basic school where lightning struck a tree behind a classroom (see Figure 9.4) and affected several children who were in class and those in the corridors. No deaths were recorded thought seven of the children fainted and some of them only gained consciousness at the hospital. Two other cases in Chibombo could not be attended to due to the constraints mentioned above. These were cases in two separate areas and incidents where in one case a man was struck to death and in another incident a woman carrying a baby on her back was struck by lightning. The woman died but the baby survived.

5. In Mumbwa only one site was visited. A man was sitting under a tree with two of his sons. Lightning struck the tree and killed the man. His sons survived with minor burns.

Western Province

In Western province, Senanga, Mongu and Kaoma were visited.

1. In Senanga town, four cases were studied in different areas with one human death. The deceased died after the house in which he was sleeping caught fire after being struck by lightning. It was not clear whether he died from lightning or from the fire caused by lightning. It is however highly suspected he could have died from lightning because efforts to wake him up before the fire grew, proved futile. The house has since been demolished because the family thinks it was an act of witchcraft.

2. Six other cases were studied in various villages. In Lanyi village a woman and her two children died on 17 October 2011, when the house in which they were sleeping caught fire after being struck by lightning. In Litongo village, a house and all household goods were burnt to ashes after the house was struck by lightning. Fortunately there was no one in the house at the time if the incident. In Royal Moyo village, a telecommunications tower and Zambia Electricity Company transformer were often struck by lightning, damaging the equipment and disrupting services. In the same village, on 2 April 2011, 15 cows were affected when lightning struck as they were grazing. Nine, (9) died while five, (5) lost consciousness. In Nanula village, lightning struck two houses, shown in Figure 9.6, affecting 15 people. All escaped with minor burns.

3. In Mongu, four cases were studied, one school incident where everyone escaped unhurt but affected in various ways. On 24 February 2012 a woman died after her house was struck by lightning. Three children including a baby escaped with minor burns.

4. In another case a tree outside a house got struck but the people in the house were not affected. The interesting part of this incident is that the entire village converged to dig out the tree (see Figure 9.8). They believed the

Figure 9.6: The People in the Two Houses in the Picture were Affected by Lightning Simultaneously in Nanula Village.

Figure 9.7: Tree on the Right was Struck by Lightning, a Steel Wire for Drying Clothes Ran from the Tree to the House. House has metal roofing.

Figure 9.8: Hole Left after Tree was Dug Out.

Figure 9.9: The Tree that was Struck by Lightning.
Several children were under the tree but only one died.

lightning went into the ground and left eggs and lightning eggs are used for good luck and protection from lightning strikes. No eggs were found.

5. In the last case lightning struck a compound killing a cow and a dog. The Mongu police also reported a case where a pregnant woman was killed under a tree after lightning struck the tree. The case could not be followed up due to distance.

6. In Kaoma, 4 cases were studied, two cases recorded a death each, one in Kaoma town and one in Kashamu village. In Kaoma town, a boy was pounding maize while his friends played around him under a tree when it started showering. Lightning struck the tree (shown in Figure 9.9) killing the boy who was pounding while all the other children survived. In another incident, nine 9 died after being struck by lightning as they were in the fields grazing in February 2012.

Northern Province

In Northern province, three towns were visited namely, Luwingu, Mpika and Kasama.

1. In Luwingu three, (3) sites were visited. Two of the sites were in Luwingu town, a police station and a Zambia Electricity Company substation. Both were struck by lightning in different incidents damaging the internet system at the police station and a 1.6 MVA, 16/11 KVA transformer of the electricity company. The transformer case was reported by the Zambia National Broadcasting Cooperation during the evening news on 1 January 2012. The police case was reported by the officer in charge at Luwingu police post when the research team went to pick a police officer to accompany them to the scenes of lightning incidents in the area.

2. The third incident was in Kapisha village where lightning struck a house killing two, (2), children out of four who were sleeping on the same mat under the same blankets. The two were at each end. The two who were in the middle survived, though they initially fainted but later gained consciousness.

3. In Mpika, a site was visited in Kopa village. Kopa village is known for two types of edible caterpillars called the yellow and the black caterpillars. People from nearby chiefdoms gather in chief Kopa's village in November for about two weeks to gather and dry the caterpillars. They camp in the bush under the trees in temporary shelters. In 2012, Lightning struck one of the trees in the camp and killed three people in one of the shelters. All the three were from the same family.

4. Two other incidents were reported where people had been killed by lightning in 2012, but the places could not be visited to verify the information.

5. In Kasama town a grade 11 pupil, male died after he was struck by lightning as he was standing under a tree. In Fwambo village, a house had been struck by lightning causing a separation in the walls. No one in the family was injured though the house was damaged. In the picture below, the man

Figure 9.10: The House where Lightning Cracked the Wall Open.

is standing in front of an open space, created when lightning cracked the attaching walls (shown in Figure 9.10) and also cracked the bedroom wall creating an opening to the outside. One of the residents remembered dreaming of a loud noise, but when he woke up in the morning, he thought he was still dreaming when he saw rays of light from the sun in the bedroom. When he went outside, he found neighbours gathered, looking at the cracks on the walls of his house.

Lusaka Province

1. In Lusaka, National Institute for Scientific and Industrial Research was struck by lightning damaging the mains switch. The institution stayed for two, (2) weeks without power and it cost the institution K30 000 000 to replace the switch. The second case was the lightning strike at the Meteorological Department of Zambia where lightning struck the UHF radio antenna mast and damaged the entire local area network equipment in the building.

Summary of the Damage Caused by Ligthning in the Cases Studies

Summary of the damage caused by ligthning in the cases sare desribed in Tables 9.1–9.3.

Table 9.1: Effects of Lightning on Humans

Location	Number of Events	Effects	
		Total Affected	Deaths
Nakonde Muchinga Province	3	Several	2
Mkushi Central Province	1	6	5
Kapiri Central Province	1	0	0
Kabwe Central Province	2	Several	2
Chibombo Central Province	1	5	0
Lusaka Lusaka Province	2	0	0
Senanga Western Province	10	Several	4
Mongu Western Province	4	Several	1
Kaoma Western Province	4	Several	2
Mumbwa Central Province	1	3	1
Luwingu Northern Province	3	Several	2
Mpika Northern Province	2	Several	3
Kasama Northern Province	2	3	1
Chipata (Not Yet Visited)	1	7	Non reported
Chadiza Eastern Province (Not Yet Visited)	1	3	2
Mpongwe Copperbelt (Not Yet Visited)	2	2	Non reported
Total	40	Several	24

Table 9.2: Effects of Lightning on Property

Location	Number of Events	Property	Effects
Nakonde	Multiple	Power poles and one ZESCO transformer	Damaged
Mkushi	Multiple	! house and Several ZESCO transformers	House burnt, transformers damaged
Kapiri Mposhi	1	Electronic equipment	Damaged
Kabwe	2	Household electrical appliances	Damaged
Senanga	Multiple	Household goods and two houses	Burnt
Lusaka	2	Office Equipment	Damaged
Mongu	Multiple	1 house and household goods	Burnt

Table 9.3: Effects of Lightning on Livestock

Location	Number of Events	Type of Animal	Effects	
			Deaths	Survivors
Senanga	1	Cattle	9	6
Mongu	1	Cattle	1	0
Kaoma	1	Cattle	9	12
Mongu	1	Dog	1	0

Discussion

As can be seen from the report, just a handful of areas out of the whole country were selected for the case studies. The team mainly followed up cases that could have been reported in the media. However, in all the places visited, the research team discovered more incidents as the people volunteered more information. It was however not possible for the team to follow up most cases due to the limitations in the programme. Judging from the statistics from the few places visited and the few cases studied, it is clear that Zambia ha a serious problem with lightning though further studies would be required to establish the seriousness of the problem.

The information collected could not be interpreted scientifically as the research team had no expertise to do so. This is a major concern as failure to interpret the information scientifically would mean no solution to the problem. Negotiations are under way to build capacity in the area of concern.

Conclusion

Lightning is a natural phenomenon and to date, there is no way known to mankind of preventing it from happening. It is therefore incumbent upon every country to address lightning safety issues. Lightning is very destructive and can be fatal when it interacts with man, livestock and manmade structures.

From the case studies, it is evident that Zambia is affected negatively by lightning in many ways. The very first rains in November, 2012, Fwambo village has recorded four deaths in two different incidents, a woman and a child in one incident and two children in the other. More studies are required to come up with a conclusion on the extent to which the country is affected.

After further studies of the physics and injury mechanisms of lightning, it will be possible to interpret the case studies and come up with scientific explanations of how lightning is affecting people, livestock, buildings and equipment. Consequently appropriate devices and systems could be designed/or adopted for protection and safety.

Acknowledgements

1. NAM S&T for exposing me to lightning protection research in Nepal, October, 2011 Symposium on lightning protection.

   2. Prof. Chandima Gomes for encouraging me to take up the research and guidance.

   3. NISIR Management for logistical and financial support during the project.

   4. Zambia Police for assigning police officers to research team during visits to incident scenes.

   5. Village Chiefs, headmen, chairmen and the individuals and groups of people who agreed to be interviewed.

   6. My Family for moral and spiritual support during the project.

References

1.   http://en.wikipedia.org/wiki/climate_of_Zambia

2.   http://geography.about.com/library/cia/blcZambia.htm

3.   http://en.wikipedia.org/wiki/Zambia

4.   http://www.google.co.uk/map of Africa showing Zambia

5.   http://en.wikipedia.org/wiki/file global _ lightning _ frequency.png

Chapter 10

Lightning in Zimbabwe: Statistics, Costs and Protection

Beaula Chipoyera

Principal Science and Technology Officer,
Ministry of Science and Technology Development,
Harare, Zimbabwe
E-mail: bchipoyera@gmail.com

Introduction

Zimbabwe is located in southern Africa. According to Sibanda (2011), Zimbabwe has a land area of 390,759 sq km (150,873 sq mi). From north to south its greatest distance is 760 km (470 mi), and from east to west it is 820 km (510 mi). The country borders Mozambique to the east and Botswana to the west. South Africa is located to the south, and the Limpopo River forms the boundary between the two countries. In the north the border is formed by the Zambezi River, beyond which is Zambia. The map in Figure 10.1 shows the location of Zimbabwe in Africa.

Zimbabwe has two major language dialects, Shona and Ndebele. Shona is the major dialect spoken by about three quarters of the Zimbabwean population which covers areas in the south, east and west of the country. The Ndebele occupy the western parts of Zimbabwe.

Lightning in Zimbabwe

Lightning is one of the major natural hazards that Zimbabwe has had to grapple with. Locally it is known as 'mheni' in Shona and úmbani'by the Ndebele. This hazard continues to cause havoc leading to deaths, injury and destruction of property and livestock. According to Johwa in website newzimbabwe.com [1], Zimbabwe is one of the world's most lightning-prone countries and is the holder of a world record

Figure 10.1: Location of Zimbabwe in Africa.

in lightning-related fatalities. Johwa further noted that the onset of the rainy season from October stretching to April brings with it a frightening phenomenon that claims dozens of lives and countless livestock. During this period it is estimated that more than 100 people perish due to lightning in different circumstances [1]. Most of these deaths are reported in rural areas owing to a number of reasons whereas in urban areas, the most affected are soccer players wearing spiked shoes in open fields such as football pitches.

Figure 10.2: Lightning Kills Hundreds in Zimbabwe Yearly: by Wilson Johwa.

Measures Recommended to Mitigate the Effects of Lightning

☆ A number of measures may be adopted to reduce the effects of lightning especially in rural areas. Some buildings, especially those with thatched roofs, are being protected by providing a 'lightning conductor' which is a very tall wooden pole at least six metres high, standing at least 1.5 metres from a building. Galvanised steel wire along the length of the pole extends beyond the end of the pole both on the top and in the earth. This tends to protect animals or children from receiving a shock during a lightning strike. If the ground is rocky or built up, lightning conductors can also be attached to trees. Where a number of houses are close together, the same lightning conductor can protect many houses [4].

☆ The empowerment of women through provision of basic facilities such as water, easier energy sources, and improved technology in farming such as use of planters, harvesters would reduce the burden shouldered by women.

☆ Government departments should work together to provide awareness programmes as well as establishment of early warning systems by the Meteorological Department and the Civil Protection Unit.

Protection against Lightning Strikes

If a person is caught up in a storm there are several actions one can take as suggested by the National Weather Services [2]:

☆ One should not lie on the ground.

☆ Stay away from sheds or open shelters

☆ Crouch in a foetal position, only vertical and stand on toes the idea being to minimise the contact area with the ground.

☆ Avoid any metal objects such as bicycles and golf clubs, fishing rods, tennis rackets or tools.

☆ Get out of the water. Water is a conductor of electricity.

☆ Spread out and do not stay in a group.

☆ If on a bicycle and lightning is within 5 miles, STOP riding, get off of your bicycle, find a ditch or other low spot and sit down.

☆ Never lie flat on the ground during a lightning storm.

☆ The ideal hiding place is a car - any car, except for soft-top convertibles. Nothing will happen to you when lightning strikes the car with you inside due to the fact that the body of the car has a wide surface, thus dissipating the electrical current and acting like a protective cage.

☆ Within a building, one should avoid anything that conducts electricity and is plugged into a wall socket (phones, electrical outlets, lights, desktop computers, televisions, stereos, and water faucets - metal plumbing conducts [3].

Reasons for High Incidences in Zimbabwe

Scientific Causes of Lightning in Zimbabwe

Zimbabwe's abundant granite outcrops may be one reason behind the country's high lightning toll. Scientific explanations are linked to the geographical formations of most parts of the country which are made up of granitic formations. According to newzimbabwe.com [1], Johwa also notes that research done by the University of Zimbabwe indicates that granite is radioactive and discharges gamma rays up to the cloud, thus ionizing the air molecules. Abundant granite outcrops, together with soot from the numerous kitchen huts, offer the much-needed opposite charge on the ground, while tall objects offer the easiest route for electrical discharges to steer its way to the ground. The dominant topographical feature of Zimbabwe is its central granite plateau, which runs diagonally from the southwest to the northeast and is covered with rich farmland. The plateau is marked by granite outcrops and hills known as kopjes and is cut by a narrow outcropping of volcanic rock that runs roughly north to south for about 520 km (about 320 mi).

Deforestation

The cause of deforestation is mainly land clearing for agriculture, uncontrolled logging, collection of fuel wood, fire, and overgrazing are also taking their toll.

Cultural Beliefs

The Shona are conscious and knowledgeable about their environment to the extent that some believe they can manipulate natural weather patterns like lightning and thunder for benevolent and malevolent factors.

Religion

Some religious groups believe that the best way to communicate with God is through worshipping in open air so that their prayers can go directly to God. Some congregate under trees without wearing shoes, hence they expose themselves to lightning during the rainy season. In November 2002, a state run newspaper the Herald reported that 10 worshippers were struck to death while 61 others sustained injuries after being burnt [3].

Nature and Possible Causes of Lightning

According to the National Weather Services, 'in simple terms, lightning is a giant spark of electricity in the atmosphere or between the atmosphere and the ground'. It is characterized by the discharge of electricity between rain clouds or between a rain cloud and the earth and is usually seen as an arc of extremely bright light which can be many kilometres in length; however, there are other forms as well. Accompanying the lightning is the giant roar of thunder. The thunder is caused by the expansion of air that has been heated by the lightning which then collides with cooler air, creating the sound of an explosion. The National Weather Service (NWS) also suggest that lightning can heat its path five times hotter than the surface of the sun [2].

The other dominant belief from an African perspective is that lightning is human induced and is directed at one's enemy as a reprisal following a conflict or a relationship gone sour with the individual concerned. Stories chronicling how lighting has been used by humans for this purpose abound, however there has not been scientific proof to that effect.

Lightning is also a phenomenon that takes place within the clouds and these tend to affect aircrafts. It can strike in the same place twice or more times because the same conditions that draw lightning are not likely to change also especially if that place features something like a high radio antenna perched on top of a mountain [6]. The weather channel reports that it is estimated that lightning hits the earth *100* times each *second* and that thus 8.6 million strikes *per* day and over three billion each year [2].

Statistics of Lightning Victims

It appears there is no credible documentation such as a database with statistics of lightning incidences. However, anecdotal evidence from newspaper reports suggest that in most cases, lightning victims will be tending livestock (mostly children) or conducting prayers in the open. Generally women are more vulnerable to lightning than men owing to the nature of the socio-cultural prescribed roles in the home. The roles that rural Zimbabwean women play in the home put them under a lot of pressure, such that they are prepared to take risks that expose them to the danger of being struck by lightning. The rural Zimbabwean women are the major source of labour in the family field, they have to fetch firewood and water and cook for the family, and thus they cannot afford to rest even when there is rain, lightning and thunder. It is when thay engage in such activities in the rain that they are exposed to the danger of being struck by lightning. Normally reports of women who fall victim to lightning are struck when collecting firewood or coming from fetching firewood or penning livestock.

Some of the chilling statistics reported in the Herald, the leading newspaper in Zimbabwe, between November and December 2012, for instance are given in Table 10.1 below:

Table 10.1: Some Lighting Reports in 2012 in Zimbabwe

Dates	Area	Event
10 Nov 2012	Alaska, Mashonaland West	5 people were killed
21 Nov 2012	Mt. Darwin, Mashonaland Central	17 people hospitalised with broken limbs
06 Dec 2012	Kezi District, Matebeleland South	4 family members were burnt beyond recognition
31 Dec 2012	Chikomba District, Mashonaland East	2 minors were burnt to death

The latest statistics reported by the Herald on Saturday 19 January 2013 is that 38 people have so far perished due to lightning strikes since the onset of the rains in October 2012.

References

1. http://www.newzimbabwe.com/pages/rain.11664.html [accessed on 18 January 2013]

2. http://www.lightningsafety.noaa.gov/science-overview.htm

3. http://www.shortnews.com/start.cfm?id=26864

1. 4.http://tilz.tearfund.org/Publications/Footsteps+4150/Footsteps+44/Protection+from+lightning.htm

5. Sibanda, F. (2011) African Blitzkrieg in Zimbabwe: Phenomenological Reflections on Shona Beliefs on Lightning. Saarbrucken: Lambert Academic Publishing GmbH and Co. KG.

6. http://starryskies.com/articles/2003/08/lightning.myth.html

Chapter 11
Lightning Incidents in Zimbabwe

Jephias Mugumbate

Senior Weather Forecaster,
Meteorological Services Department of Zimbabwe,
Harare, Zimbabwe
E–mail: jephbatsi@tsamail.ca.za

Introduction

Lightning is an atmospheric discharge of electricity that is frequently accompanied by thunder when it occurs during a thunderstorm. It is electricity that is discharged from a cloud. According to Colin Buckle in Weather and Climate in Africa (1996), lightning is the sudden flash that accompanies discharge of atmospheric static electricity as it overcomes the normal resistance of air.

It can also occur during volcanic eruptions and dust storms. The discharge produces a gigantic spark, the lightning flash between oppositely charged parts of a cloud or from cloud to another cloud or between a cloud and ground. The front end of a bolt of lightning can travel 60,000 miles per second and reach 30 000 degrees Celsius.

An average bolt of negative lightning carries a current of 30,000 amps. It rapidly heats the air in its immediate vicinity to about 20,000 degrees Celsius. This compresses the surrounding clear air and creates a supersonic shock wave that decays and becomes an auditory wave we call thunder. Lightning often strikes more than 3 miles from the thunderstorm far outside the rain or even thunderstorm cloud.

In the African tradition there is a general belief that lightning can be manufactured by people. It is also believed that this phenomenon exists and is man made with the intention of killing.

Most of the lightning victims in Zimbabwe are peasants living in thatched huts in rural areas struck mostly when sitting or sleeping on the ground in their huts. At the beginning of the rainy season the storms are violent due to high temperatures that are experienced this time of the year. This is also the time when severe weather such as strong winds, flash floods and heat waves occur in Zimbabwe.

Lightning deaths have also been reported in urban areas which can be attributed to urban heat islands due to intensive construction as well as emission of gases from green houses. The bulk of Zimbabwe has granite outcrops that also contribute to the high incidences of lightning strikes. Some academics such as Max Van Olst from the University of Zimbabwe's Electrical Engineering Department in 1987 claimed that Zimbabwe's fertile soils play a role in causing lightning deaths. He argued that the Zimbabwean soils are poor conductors of electricity which forces the charge from lightning to stream with concentrated force up to hundred metres from the strike point as it follows a narrow path of easy conductivity instead of dispersing evenly at the point of the strike.

Evidence of Lightning Incidents in Zimbabwe

The onset of the rainy season revives flora and fauna. While for the plants it's time to sprout, blossom, bear seed, beautifying the earth, turning it green and providing pastures to living organisms. For human beings however, the season is accompanied by this life threatening natural phenomenon known as lightning. It has claimed human and animal life and even vegetation. Zimbabwe is among countries which has highest lightning deaths incidences in the world. This was recorded in the Guinness Book of World Records in December 1975 after killing 21 people in the Eastern Highlands village of Chinamasa.

The Eastern Highlands, parts of the western districts and areas in the north east of the Zambezi valley have high incidences of lightning. This is due to the high altitude, high humidity and high temperatures. These factors combine and affect the density of the air making the areas more susceptible to lightning.

During the rainy seasons, the newspapers are awash with reports on lightning strike incidents. For example the local Zimbabwean Herald of the 16th of January 2006 reported that10 people were killed during an open prayer session in Seke Chitungwiza in 2002. The same paper reported that 13 head of cattle which were struck and killed by lightning in Msengezi District on the 8 March 2012. On the 19th of January 2013 the same paper reported that lightning claims 124 lives. Many incidences could have gone unreported.

Role of Civil Protection Unit in Reducing Lightning Deaths Incidences in Zimbabwe

The government of Zimbabwe has put in place the Civil Protection Unit (CPU) whose responsibility is disaster management issues, lightning included. It coordinates multi-sectors and collaboration of the Meteorological Services Department, Agritex, Ministry of Health and the Police just to mention a few. The unit visits communities, mostly in rural areas, to train and educate the citizens on disaster risk reduction, response and prevention and capacity assessments.

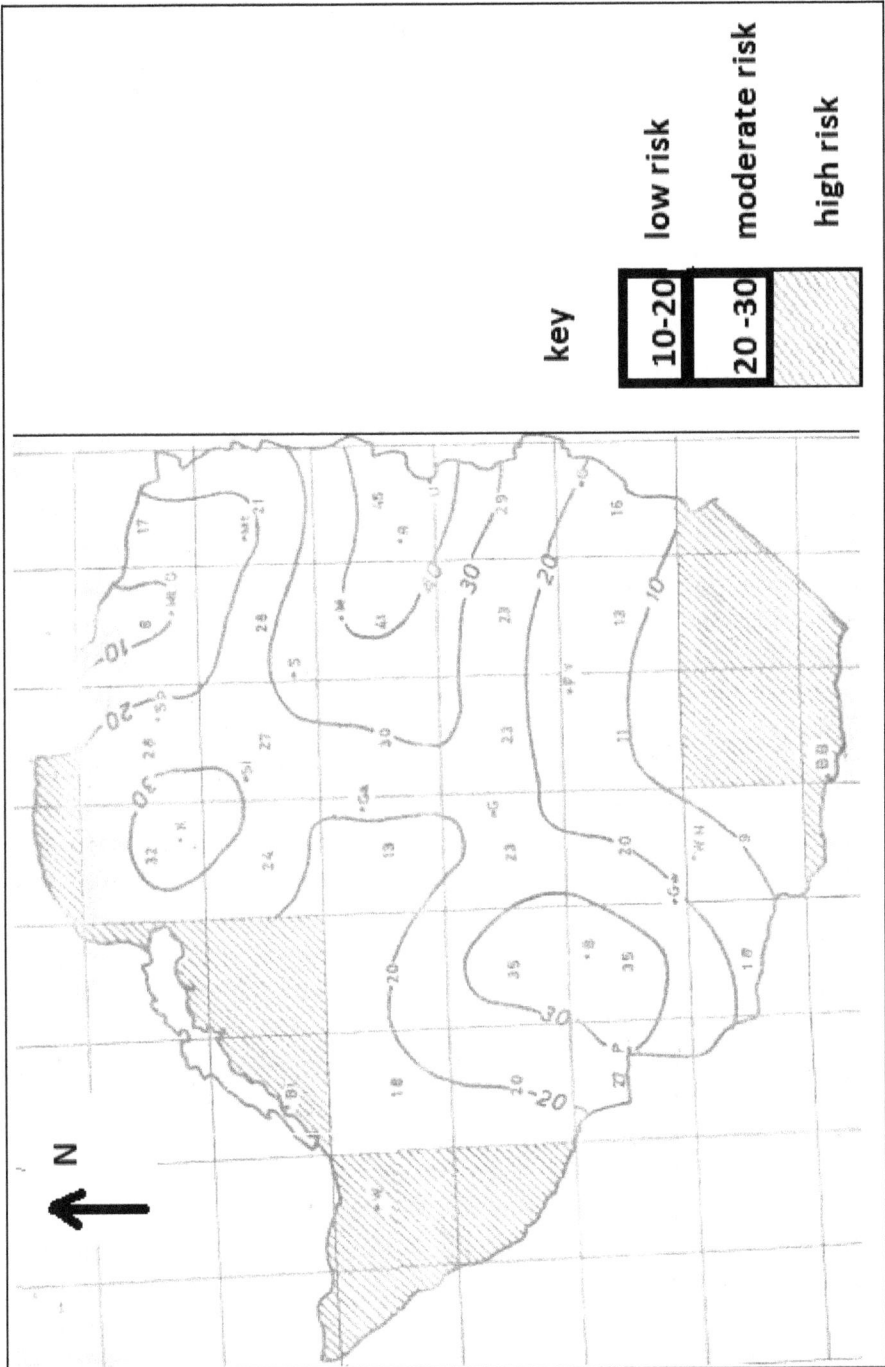

Figure 11.1: Show the Zimbabwe Lightning Risk Map.

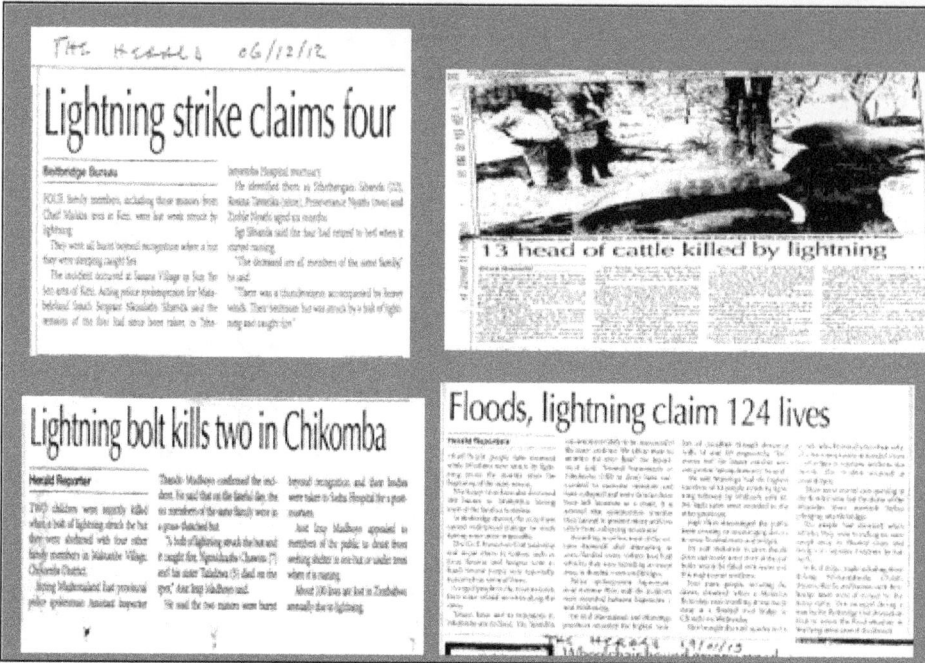

Figure 11.2: Some of the Headings that Appeared in the Local Newspapers.

The CPU educates the general public on what to do to avoid being struck by lightning. These include alerting people on the dangers of switching on their electrical appliances during a thunderstorm. They encourage them to avoid holding metal objects and not to work in their open fields during storms. Other issues that are taught are to avoid swimming as well as not to stand under tall objects like towers in a thunderstorm. The CPU encourages everyone to lie in a ditch if possible when caught in an open field during lightning storm. People are encouraged to install lightning conductors on their houses.

However, despite the vast training and warnings issued regarding severe weather, the number of communities affected by lightning is on the increase. This may be attributed to lack of resources for installing lightning conductors especially in the rural poor communities. Among other reasons are the extreme weather events due to climate change. At the same time awareness is not yet at the desired level due to lack of resources on the part of the CPU. So much folklore also exists about lightning hampering the scientific advices that have been issued as they are ignored by some people.

Conclusion

Lightning has been observed to cause lot of damage to both life and property, hence the need to combat it. The nation should continue planting more trees. Awareness programmes should be held in all areas. Cheaper lightning conductors must be manufactured. Lightning detectors and weather radars for monitoring severe storms needs to be installed.

References

1. Colin Buckcle (1996); Weather and Climate in Africa.

2. United Nations (2010); Natural Hazards Unnatural Disasters.

3. Ailsa Holloway (1999); Risk Sustainable Development and Disasters Southern Africa.

4. CHB Publication (1995); Earthing and Lightning Consultants Handbook Publication.

5. Zimbabwean Newspapers.

www.ingramcontent.com/pod-product-compliance
Lightning Source LLC
Chambersburg PA
CBHW021433180326
41458CB00001B/260